Recreational Vehicles
Second Edition

Finding the Best Buy

Bill Alderman, Jr.
Eleanore Wilson

Bonus Books, Inc., Chicago

00 99 98 97 96 5 4 3 2 1

Library of Congress Catalog Card Number: 96-85975

International Standard Book Number: 1-56625-070-6

Bonus Books, Inc.
160 East Illinois Street
Chicago, IL 60611

Cover design by Brian Malovany

Composition by Point West, Inc., Carol Stream, IL
Printed in the United States of America

Contents

Acknowledgments

We wish to thank Bill's wife Jenny for all of the initial editorial work and her encouragement and patience; June and William Alderman Sr. for their seasoned camper's advice; Maxine and Bill Conroy, longtime RV dealers and our teachers over the years; Jerry and Ann Leetzow, wonderful friends for longer than anyone cares to remember; Dean Thompson, always a leading salesman and RV expert; Alice Gonnerman, first-time camper and editor with many useful suggestions; Bob Holloway, Ph.D., a guy we can count on for good advice, honest critiques, and support; Dr. Lloyd B. Jensen, author of many textbooks; and last but not least, Eleanore's father, Denver Orvis, experienced camper, critic, and her best friend.

Prologue

Since this book was written, a number of new RV models have appeared on the market and the few that have disappeared remain in the used market. The newer models are sleeker looking, often better designed with wider bodies, up from 96" to 102", and slide-out rooms are available in everything from fold-downs to class A motor homes. (See the park model floorplan on page 30 for an example of a slide-out.) The price ranges outlined in chapter 6 have increased but are still relevant when comparing one category with another. For example, fold-downs are still the least expensive with class A motor homes topping the list. The buying sequences suggested in this book and the checklists and common-sense admonitions are still relevant. So are the campsite tips. And the used RV advice is perhaps more important now than ever before, since that market is expanding.

In any case, good luck with your new or used RV, and smooth driving wherever you go.

Bill Alderman, Jr.
Eleanore Wilson
July 1996

Introduction

This book is dedicated to people who are giving serious thought to buying a recreational vehicle (RV). It is dedicated to those who have the fear of spending money on something they are not certain they will like and to those who have such questions as Should I buy new or used? Should I buy large or small? Should I buy a motor home or a travel trailer?

The authors are about to share a combined 35 years of camping experience and 11 years of experience in selling RVs. We hope that this book saves readers time and money and helps them spot potential problems.

We suggest that the whole book be read before any action is taken on the shopping sequences recommended. Readers may decide to rearrange the steps for their own convenience.

We've been there, too. And we have never forgotten our needs and the urgent questions we had.

The contents of this book represent the opinions and personal experiences of the authors and do not necessarily represent the viewpoints of the RV industry in general.

Is RV Camping for You?

Chapter 1

Picture this: You are wearing a light jacket and sitting in a comfortable lawn chair under an awning. The time is midnight, and you are reflecting on a pleasant day of sightseeing, hiking, fishing, or of visiting old friends. A slight breeze carries the aroma of a pine forest mixed with the fragrance of a camp fire still flickering a few feet away from your chair. Somewhere an owl hoots, and everywhere crickets are chirping. The sky is filled with a million stars, and as you peer through a gathering ground mist, you can just make out the outline of the surrounding mountains in the moonlight. Not 40 feet away, a stream gurgles its way down the mountainside. The picture is peaceful, it's paradise, and it's waiting for you to stop imagining it and to get into it.

Missing are an alarm clock and a telephone, luggage to haul in and out of a motel, lights that flash across the window and trucks that roar by, and a room in which 40-watt bulbs shield you from the truth about its cleanliness.

All of that is missing because you have gone camping in an RV. People don't know where you are unless you decide to tell them. For a time, you are in your own portable world, with no schedules to tie you down. Your home on wheels can contain as many of the comforts of home as you choose to include, with the possible exception of a water bed.

If this all sounds enticing, the chances are great that you will or do enjoy camping. (If you are an experienced camper and know that you love it, skip to chapter 2). This is *the* most important question that you, if you are a first-time camper, must ask yourself. If you have a family to consider, it is also important that they agree. Nothing can spoil a trip faster than someone noisily longing for a motel room.

Still not certain that you will like camping? Rent an RV for a weekend. (Large travel trailers usually are not available for rental because they require a heavy-duty, frame-mounted hitch on the tow vehicle.) Rental isn't inexpensive, but then neither is losing money on the resale of an RV that you purchased before you discovered that you dislike camping. Plan to spend some time and, yes, some money answering this question. By the way, if you rent, rent a unit that is large enough to accommodate comfortably the number of people who will occupy it. Keep in mind that you may have a rainy day and that you do not want people tripping over one another because they have to spend some time indoors. A too-small unit can ruin an otherwise good time, not to mention affect the condition of your family's feet, shins, and elbows.

Rent from a reputable dealer. Ask questions about the maintenance program, and ask if you will get instructions for operating the installed options and appliances. If you do not like the answers, find another dealer. Also check on insurance coverage and who is responsible if something should go wrong. Then get set to enjoy your first RV experience.

The Many Faces of RVs

Chapter 2

The rest of this book is going to assume that you are seriously considering buying an RV or have already decided to do so. We will take you step by step through the questions you should be asking yourself and others. The first question in the "buy" category should be, What type of RV is right for me? A good place to start is with a review and understanding of the choices. A quick overview of various types of available RVs follows.

What's Out There

Trailers

Fold-down. A popular choice for those on a low budget or those who need an RV that can be garaged and is light in weight.

Travel trailer. This is usually selected when the buyer is looking for more creature comforts when camping, coupled with the flexibility of having a vehicle to drive while leaving the trailer at the campsite.

Park model. Designed to be parked semipermanently at a campsite and used, for example, as a summer house.

Fifth-wheel. Preferred by people who demand all the comforts of home when on a trip and who believe the fifth-wheel trailer is easier to tow than a travel trailer.

Motor Homes

Van camper. Great for the small family that travels light and is looking for an economical, compact, and comfortable traveling machine.

Mini motor home. A good compromise of size and economy between a van or micromini and a class A motor home (see below).

Micromini motor home. Usually provides more living space than the van camper; at the same time, it offers economy of operation, having less bulk than the standard mini.

Class A motor home. People who need lots of space for both storage and living and who enjoy a view over the traffic ahead will choose the class A.

Truck Campers

These vehicles are for those who own or need a pickup truck, who enjoy roughing it a bit when camping, and who do not wish to buy either another motorized vehicle or a trailer.

Wheel Configurations

In your analysis of the similarities and differences among various types of recreation vehicles, you will note that there are several styles of wheel configurations. Three of these arrangements, along with the correct terminology for them, are as follows:

Tandem wheels. (figure 2-1)

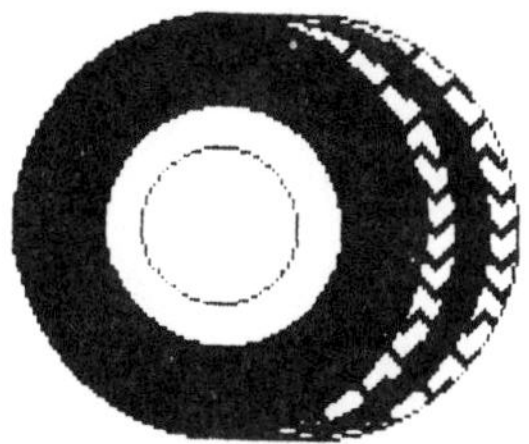

Dual wheels. (figure 2-2)

Tandem wheels. One set of wheels is behind the other. You will find these on trailers (see figure 2-1).

Dual wheels. Two wheels are side by side on each end of a single axle. These are found on motor homes and on some tow vehicles (figure 2-2).

Tag axle. A rear axle has dual wheels, while another axle right behind it has single wheels. This configuration is used to cut down on overhang of the RV and to support the extra weight of a long motor home (figure 2-3).

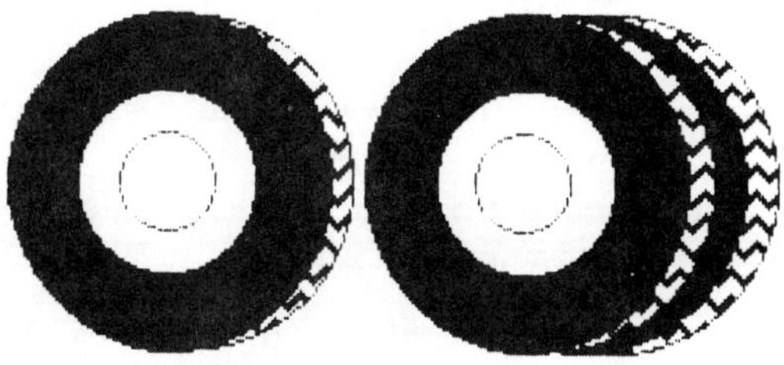

Tag axle. (figure 2-3)

Detailed descriptions of variations on the three vehicle types named above follow.

Trailers

Fold-Down, Pop-Up, Camping, or Tent Trailer

Any one of the terms can be used interchangeably. However, *camping*, or *tent trailer* is more often used in the RV industry. We will use *fold-down* in this book, as it seems somehow less ambiguous to us. From the side, they look something like figures 2-4 and 2-5.

These trailers can be an especially good choice for someone just starting out without a lot of money to spend. Seasoned campers also consider fold-downs when they grow tired of sleeping on the ground in a tent but they aren't quite ready to give up the "feeling" of tent camping. Fold-downs do give you the tenting experience, while at the same time providing many of the comforts of home.

Advantages

- They are small, compact, and can be garaged.
- They are light enough in weight so that they can be towed with some of the new smaller cars.
- Gas mileage from the tow vehicle is good because of decreased wind resistance.
- Most are easy to set up; however, if you have any health problems, remember that most of these units have to be *hand cranked* from the folded position to the "up" position at the campsite.
- Some of the larger fold-downs offer such options as toilet and shower, with holding tanks for wastewater; furnaces; air-conditioning; and, yes, the inevitable TV

Fold-down trailer, closed for towing. (figure 2-4)

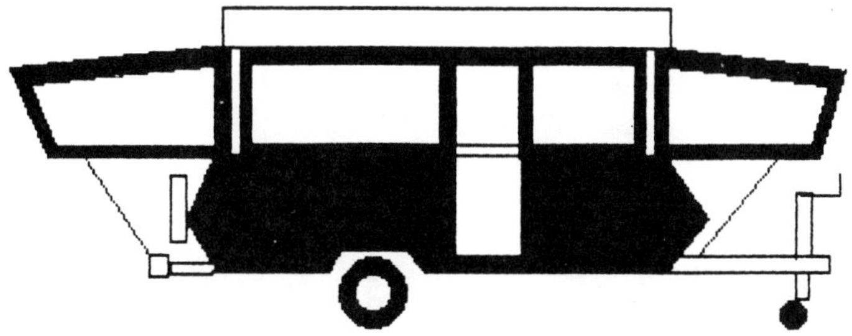

Fold-down trailer, opened at campsite. (figure 2-5)

antenna. So you can make this trailer as much like home as you wish.

Disadvantages

- Canvas-sided models tend to flap in a storm or high wind. This can be annoying to some people, but some might consider it an advantage.
- Touching the inside of the canvas during a rainstorm can, from oil in the hands, cause water seepage at the spot touched. A good reason to keep the fabric in good condition and waterproof it periodically.
- The canvas- or vinyl-walled tent areas are not always insect-proof where they join to the body of the trailer.
- Folding the unit down in the rain means that when you arrive home it must be opened up again to allow the

fabric to dry thoroughly. If it isn't, mildew will result. Setting up one of these trailers in the rain can also be "fun." Try to do it without getting the mattresses wet in the extending bed ends. You have to be very quick! This is an inconvenience to some people.

- Using any of the appliances or the toilet during brief stops on the road is very inconvenient at best, if not impossible, unless, of course, you don't mind cranking it up and down at the rest areas.

Fold-downs are also available with hard sides. They usually take a little longer to set up, and they are sometimes heavier. However, they do eliminate some of the disadvantages of canvas in a rainstorm, such as the flapping and waterproofing worries.

Some manufacturers use vinyl fabric instead of canvas. With these trailers, you don't have to worry about touching the sidewalls in a rainstorm. If mildew does occur, it is easier to clean off. However, the rest of the disadvantages, as well as all of the advantages, still apply.

Travel Trailers

For the uninitiated, travel trailers are *not* referred to as pull-alongs, mobile homes, or tow-behinds. They are travel trailers, and their silhouette looks like figure 2-6.

People select travel trailers for several different reasons. Some are looking for a unit to park permanently on a lot that they lease by the year from a campground or one that they have purchased. Others, looking to travel with their trailer, want all of the comforts of home coupled with minimized setup time. In addition, they want the convenience of having the tow vehicle available for short trips. Most travel trailers are self-contained; that is, they have a shower, toilet, and holding tanks for wastewater.

Travel trailer. (figure 2-6)

Advantages

- Already mentioned but worth repeating here: All the comforts of home are available with most travel trailers.
- Unlike most fold-downs, most units are well insulated from floor to ceiling.
- They are easy to set up at the campsite.
- The trailer can be left at the campsite while the tow vehicle is used to run errands or for short side trips.
- If the tow vehicle or trailer should break down, the other unit is still available for use while repairs are made. Also, the tow vehicle can be replaced without the expense of replacing the living quarters (as in the case of replacing a motor home).

Disadvantages

- There is no escaping the fact that trailers can be intimidating to haul around. Perhaps you would rather not tow something down the road that is more than likely larger than the tow vehicle itself. Most people get used to the idea quickly and make the necessary adjustments in their driving habits, but some do not like it.
- Because travel trailers are higher off the ground than a fold-down trailer, wind resistance is greater. This can cut gas mileage by as much as 50 percent, and you are

more vulnerable to crosswinds and large trucks passing on the highway.

- Hitching and unhitching in bad weather can be a pain; however, a power tongue jack, available from the dealer as an after market option, can eliminate the cranking.
- Backing up into a campsite is a problem for some people (we will provide some tips later in this book that make this job easier).
- If you want to use the facilities of the trailer, you must leave the tow vehicle and walk outside to the trailer—an inconvenience (although a minor one) in foul weather.

Park Model Trailers

Unlike travel trailers, these units do *not* have holding tanks for wastewater. They are built to be parked on a lot and seldom moved except to change to another "permanent" location with full hookups for sewer, water, and electricity. Many park models offer a "tip-out" or "slide-out" option, meaning that a portion of the living room and/or bedroom wall will tip or slide out from the side of the trailer to expand the living area, much like a room addition. A few travel trailers and motor homes also offer this feature. Mostly, they are found in park models because the added setup time required is not something the average camper wants to do when traveling. Frequently, these vehicles include a sliding patio door. Park models are not pictured here, but they resemble travel trailers.

Advantages

- Park model trailers usually feature full-size home appliances. Bathrooms usually will have a full-size tub and toilet.

- With a slide-out, they are roomier per linear foot of trailer than a travel trailer without this option.
- Park models usually cost less per linear foot than travel trailers of comparable quality. The reason for this is the lack of holding tanks and the fact that standard household appliances are cheaper than those built for travel vehicles. Travel trailer refrigerators have two, and sometimes three, modes of operation (gas, electric, or battery).

Disadvantages

- If the unit is parked on a semipermanent basis on a lot, there are no disadvantages. If it is used as a travel trailer, all of the above advantages become disadvantages.

Fifth-Wheel Trailers

This term is really a misnomer. These trailers have a hitch point just like a travel trailer (see figure 2-7). The name probably came from the fact that these trailers usually have tandem wheels (see above); that is, they have four wheels, one pair in front of the other, with the "fifth wheel" being the hitch point. Fifth-wheel trailers are goosenecked in shape and are designed to be towed either by a pickup truck or a customized vehicle made for the purpose.

Advantages

- The advantages are the same as for travel trailers, but with two additions: Some claim that a fifth-wheel trailer is easier to tow and more stable on the road. On the other hand, others who have towed both types don't see any difference if the travel trailer's tow vehicle is properly equipped and has an appropriate hitch with sway control.

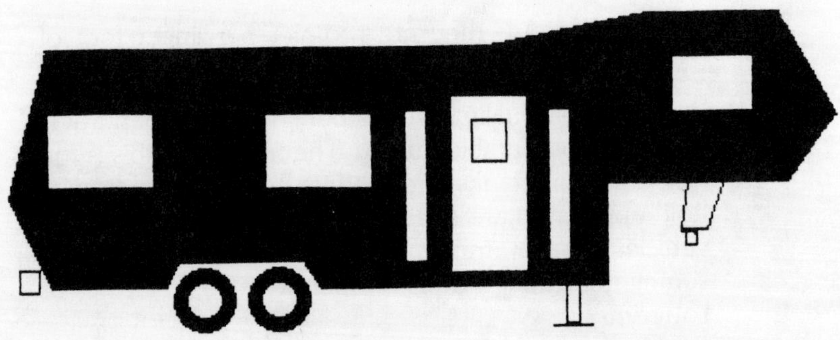

Fifth-wheel trailer. (figure 2-7)

- Because the combination of tow vehicle and fifth-wheel trailer more closely resembles a semitrailer truck, it has a shorter turning radius than a travel trailer and therefore is more maneuverable in tight parking situations.

Disadvantages

- Again, disadvantages of the fifth-wheel trailer are the same as for the travel trailer. However, in most fifth-wheel trailers, you cannot stand erectly in what is usually the bedroom area, which is situated over the bed of the pickup truck. It can be argued that you don't need to stand up if you are going to bed, but this is a disadvantage for those who like to do some stretching while standing. Some manufacturers now offer a stand-up bedroom by making the neck of the trailer higher or by placing the living room where the bedroom is usually located. If the latter option is selected, the problem of stand-up space is transferred to the living room.

- Some people do not like the idea of riding around in a pickup truck or have families that are too large to fit in the cab while traveling. For the latter, crew cab trucks are available that provide the extra seating capacity, but they usually carry a higher price tag.

Motor Homes

A word about motor homes in general. They are *not* mobile homes, which are large live-in trailers found in mobile home parks. Four basic types are available: van camper, mini, micromini, and class A. For the most part, these units are self-contained, as defined earlier. Unlike trailers, they have their own engine and can be driven like a car. Each type is described with its own advantages and disadvantages:

Van Campers

These vehicles are what their name implies: They are vans that have been adapted or converted for camping purposes. Some offer permanent raised roofs; others offer pop-up roofs that can be raised at the campsite for stand-up space. Still others have the standard height van roof, which requires that you bend slightly when moving through the vehicle. One manufacturer lowered the floor in certain areas of the vehicle to provide stand-up space. Usually this is done in the kitchen area and at the expense of ground clearance.

Van campers also offer a choice of body styles, including

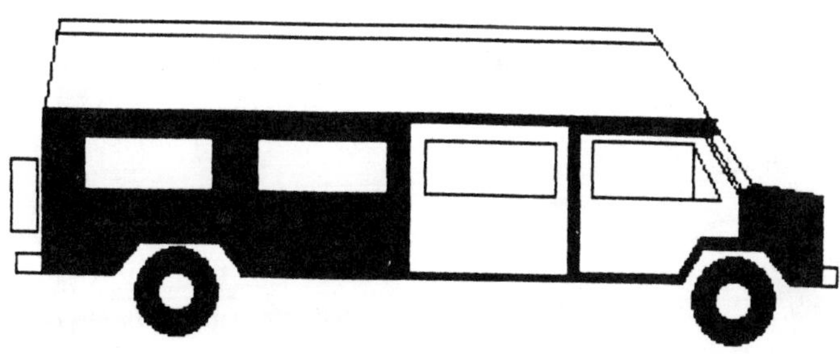

Van camper, standard. (figure 2-8)

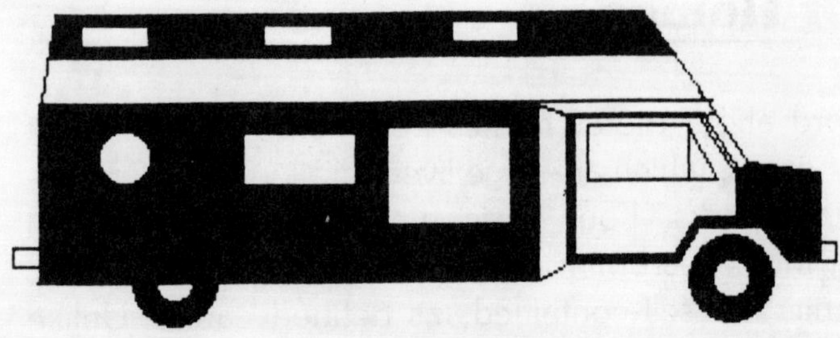

Wide-body van camper. (figure 2-9)

standard vans both foreign and domestic, that have been converted, and vehicles known as wide-body vans (figures 2-8 and 2-9). The wide-body vans offer more interior space than the standard models.

Advantages

- Because they are small and compact, vans can be used as second family vehicles when not being used for traveling. They provide good gas economy as well.
- Most vans offer toilets and/or showers as an option so that, along with kitchen and sleeping facilities, they are self-contained. As with all motor home style units, you do not have to leave the vehicle to use the onboard facilities.
- Because of the van's small size, they can be fitted into camping spots that would not accommodate larger units. Setup time at the campsite is minimal.

Disadvantages

- Van campers usually are set up to sleep at least four. If there will be four of you on a long trip, you had better like each other a whole lot, especially on rainy days.
- Because van campers as a class lack large storage areas, less clothes and food supplies can be carried than

in a larger unit. That translates to more frequent stops at laundromats and grocery stores.

- Resale of the vehicle may be more difficult than a larger coach, partly due to higher mileage than on a larger motor home that probably would not be used as a second car. As you can see, what is a decided advantage for some people can be a two-edged sword.

Mini Motor Homes

These units are also called "class C" motor homes because of the type of chassis on which the body is mounted. Class C is a van chassis that usually has been beefed up and extended in order to take the extra weight and size of the motor home body. The front end of a mini looks like a van, but in most cases has an overhang above the cab to accommodate a double bed (figure 2-10). Most come with dual rear wheels for stability and load-carrying capacity.

Advantages

- The mini offers a good compromise between the micromini (described next) and the big class A motor home in terms of overall size, interior space, and gas economy. Sizes range from 18 to 26 feet, with some companies making 28-foot models with tag axles. As noted earlier, a tag axle is another set of wheels mounted directly behind the dual rear wheels to provide for the extra weight and also to cut down on the length of the body overhang in back of the wheels. Interior space allows for full stand-up head room, along with ample storage capacity and all the comforts of home. Gas economy is, on average, between 9 and 12 mpg.
- If it is not used for sleeping, the double bed over the cab makes an excellent storage area. In addition, both

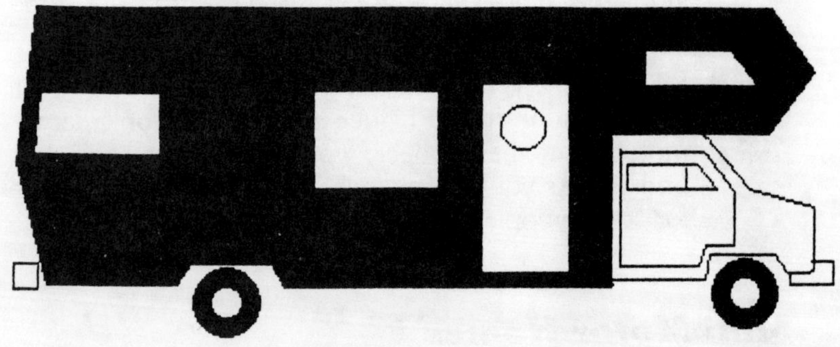

Mini motor home. (figure 2-10)

the driver and passenger have their own exit doors. Such exit doors are options on class As, which will be described later. They are more a matter of convenience rather than a safety factor inasmuch as the windows in most motor homes are large enough to be used as emergency exits.

Disadvantages

- Working on the engine is difficult because of its position in relation to the hood and inside engine cover.
- For extended trips, in comparison with the class A motor home, the mini usually has fewer outside storage compartment areas and often will not have as smooth a ride.

Micromini Motor Homes

As you might guess, microminis look like scaled-down minis. The camper body is usually mounted on a small foreign pickup truck chassis, although some manufacturers use a small domestic pickup chassis. Like mini motor homes, they have a double bed mounted over the cab. Microminis range from 18 to 21 feet in length. Most come with dual wheels. Because they look like a mini, a picture is not included here.

Advantages

- Economy! You can expect cruising mileage in the 16- to 20-mpg range.
- Considering their size, the ride usually is exceptionally good.
- Stand-up room is an advantage over some vans that do not have a raised roof.
- They can be equipped like a full-size motor home.

Disadvantages

- Because most microminis have a four-cylinder engine, they are judged by most owners as underpowered for mountain travel. This is the price you pay for economy. You must shift to a lower gear and spend most of your time in the truck lanes. Almost all of the older models have a four-speed manual transmission and no dashboard air-conditioning. Both automatic transmission and dash air-conditioning are available in some newer models.
- Because they are smaller than most minis, there is less storage space and less interior room. Passage from the cab to the interior of the motor home is more difficult than in a mini or van camper.
- Because the chassis can support less weight than a mini, lighter construction is used in order to stay within the safety limits of the gross vehicle weight. *Quality* of construction, therefore, is more important. Molded fiberglass bodies should be considered in purchasing these vehicles. These bodies have fewer body seams that can unseal and leak due to flexing while on the road.
- Cab rust is a problem if they are not rustproofed.

Class A Motor Homes

The term *class A*, like *class C*, refers to the kind of chassis on which the camper body rests. Class A chassis are full heavy truck types or, in some cases, bus types. They range in length from 21 to 40 feet. Of the motor home classifications, they are the most roomy and contain the most storage space. A class A motor home is shown in figure 2-11.

Advantages

- In the larger sizes, class As shine on extended trips because of their openness, generous storage areas, and spaciousness. There is nothing like moving-around space on a one- or two-month trip. When swiveled to face the rear, the cab seats become part of the living area. You can do this with some minis, too, but the open feeling is not there.
- The quality of the ride is superior and handling is great considering the size of these units.
- Driving visibility is excellent because the driver is sitting high above the traffic.

Disadvantages

- Some people fear driving something that appears so large. However, with a little experience, most learn to love driving the class A.
- Gas mileage is low. You can expect from 5 to 9 mpg—but you'll get 9 mpg only when coasting downhill with a tail wind! On the other hand, you would get this same mileage towing a large travel trailer with a big-engine tow vehicle.
- Remote campsites or small camp spaces may not accommodate the longer motor homes.

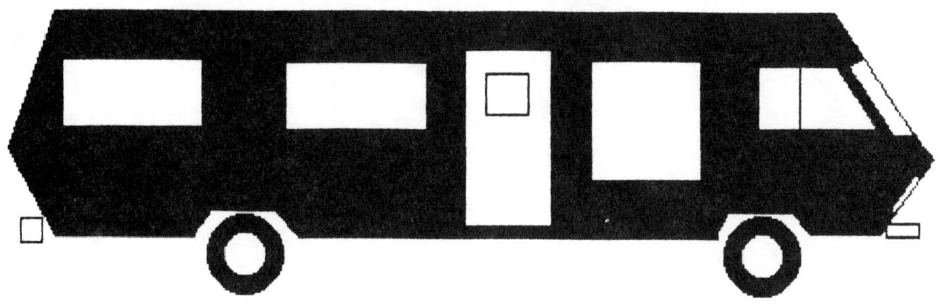

Class A motor home. (figure 2-11)

Truck Campers

These units are what the name implies: They slide into the bed of a pickup truck, giving the whole unit the appearance of a mini motor home. Another term for the truck camper is a *slide-in*. Truck campers are available in different lengths, depending upon the length of the truck bed and how far they hang over the back. Some come with full baths. Because truck campers are attached to a truck and do not have their own wheels, they are not required to be licensed or titled in most states. Figure 2-12 shows how they look off the truck.

Advantages

- By adding options, you can make these units as homey as you like.
- When you return from camping, you can remove the camper from the truck bed, and presto! you have your pickup truck back.
- They are very compact. Because they do not extend much beyond the bed of the truck, rear overhang is minimal.

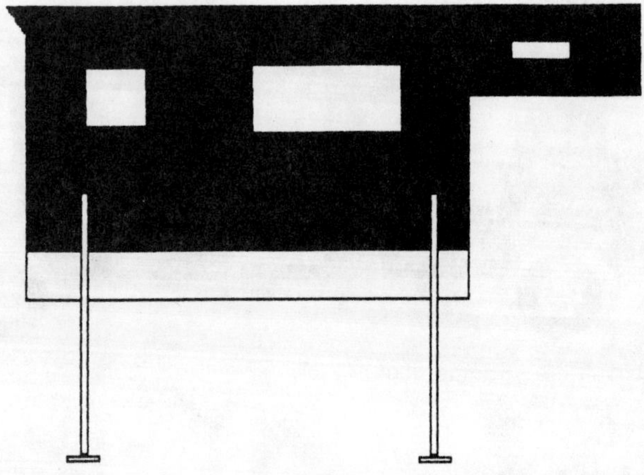

Truck camper, shown off the truck. (figure 2-12)

- Some of these units are available with a roof that can be popped up on arrival at a campsite. This gives the advantage of low wind resistance while traveling, thus better gas mileage, as well as stand-up room while camping.

Disadvantages

- Lowering the jacks to remove the unit from the truck is not easy.
- Floor space is limited to the area of the bed of the truck plus any overhang. Storage space suffers as well.
- Because the center of gravity is raised by the truck/camper combination, truck campers tend not to be as stable on the road as other types of motor homes.
- Resale is poor in certain parts of the country where not many pickup trucks are found. Also, the market for them is limited to very small families, as most pickup trucks are not of the crew cab type.

O.K., we have just dumped a lot of information on you. Now, what do you do with it? Well, you can start by reviewing the disadvantages and crossing off the units that do not fit your idea of a way to camp. For example, you may not want to drive a pickup truck. Right off the bat, you can eliminate truck campers. Perhaps you don't want to be always hitching and unhitching. You just eliminated any type of camping outfit but motor homes.

Keep going through the important factors, giving each unit careful consideration. After narrowing the field, begin looking at the *advantages* of the ones remaining on your list. In your opinion, if the advantages of certain types seem to outweigh the disadvantages, place those units at the top of your list.

A couple of additional suggestions are in order at this point. Your selection should be based heavily on the kind of traveling you plan to be doing.

- If you and your family are sightseers and plan on frequent changes of campgrounds, consider motor homes or vans as a strong possibility because they are the easiest to set up at the campsite.
- If, on the other hand, you plan to spend a week here and a week there, think about trailering, because you won't be setting up camp each night and having to hitch and unhitch. In addition, you will have a vehicle to use for short side trips or errands without having to tow the camper around with you.

Narrow the field as much as possible by moving carefully through this process. Then go on to the next chapter. It deals with the floor plans and size choices available that will help you further refine your options.

Floor Plans and Size of Unit

Chapter 3

Choosing a floor plan that most closely meets your needs is one of the most critical decisions you will make in purchasing an RV. A decision here will also help to determine the type of unit to buy, because the floor plan is in some cases interrelated with type. Obviously, if you decide that you cannot live without a permanent double bed, a permanent dinette and a full living room, don't bother looking at vans, fold-downs, or truck campers. Nor can these needs be met in a 23-foot motor home. So your selection of floor plan also will have a direct bearing on the size of the unit.

Because it is not possible to cover all of the plans available on the market in these pages, we will pose several critical issues for your consideration, with the next step being your visiting an RV show in your area or shopping at dealer showrooms. This is the best way to become familiar with the huge variety of floor plans. To give a general idea of the range of choices, a few example plans are included at the end of this chapter.

Terminology of Floor Plans

Before getting into descriptions of various floor plan options, it would be useful to run through a list of definitions of elements that will be mentioned by sales representatives and will be part of the language of descriptive brochures. Here are the features you will encounter most frequently:

Gaucho. It looks like a couch but isn't. Actually, it is two or more thick foam cushions, one of which forms a back rest and the other a seat (figure 3-1). The seat cushion typically lies on a plywood base above a storage area. The plywood sometimes has legs attached to the front so that it can slide out away from the wall, allowing the back rest to lie flat, thus forming a double bed. Gauchos are firm, yet comfortable. Try one out at a show or at a dealer.

Couch. A couch differs from a gaucho in that it has some form of springing under the foam or padding to provide seating comfort comparable to the living room sofa in your home. Most of them make up into a double bed. If you become interested in a particular unit, have this feature demonstrated. Some of these models actually are a sofa-bed complete with mattress. We don't picture one here because almost everyone knows how they look and how they are to sit on.

Dinette. This feature consists of two facing benches with a table between, much like a booth in a restaurant. The difference is that this booth can be made into a bed. Most dinettes have storage areas under the cushions, but don't assume this—check! Under these seats is a favorite place for the designers to hide water tanks, a hot water heater, or even a furnace. So lift the base under the cushions and look.

Permanent bed. This bed, double or twin, is similar to the ones at home and cannot be folded away. It usually

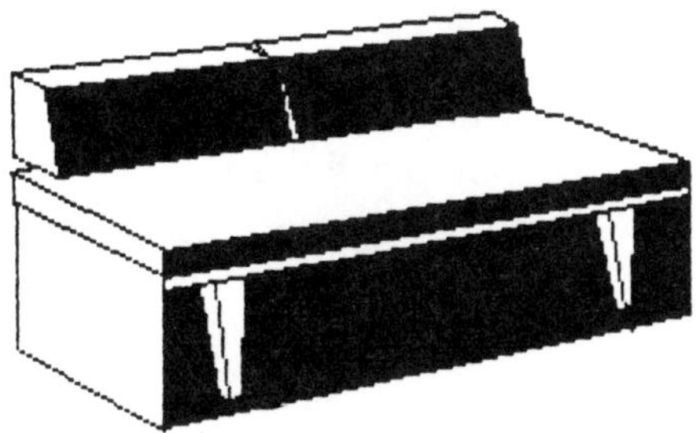

A gaucho. This one has legs for support when it is opened into a bed. (figure 3-1)

will have a thick enough mattress, but unlike the beds at home, this mattress often rests on a piece of plywood. Again, check it for comfort.

Bunk beds. These also are usually permanent beds, but they are twin beds stacked one on top of the other. They can be found in motor homes and travel trailers. Located in a separate area of the unit, bunks are ideal for children, who can go to bed while adults still occupy the living area.

Cabinet bunks. As the name implies, these bunks are cabinets that can be folded down to form beds suitable for small children. They are usually designed to support maximum weights of 100 to 150 pounds.

Wet bath. Imagine moving your home bathroom lavatory and toilet into the shower. Yes, everything would get wet when the shower is used, and that's why this arrangement is called a wet bath. Wet baths might not be for everyone, but they do save space, sometimes enough for an extra clothes closet. They are usually located at the center of one of the unit's sidewalls.

Dry bath. Just like the one at home: separate toilet, lavatory, and tub with shower. They do take up more space than the wet bath. Deciding between the two is strictly a matter of personal preference. Dry baths can be located at or near the center of one sidewall of the unit, but they are most commonly at the rear. Split baths are normally a feature of larger RVs. Commonly located between the kitchen and rear bedroom, they place the tub and shower on one side of the aisle and the toilet and lavatory on the other. Splitting the bath or placing the bath at the center of the sidewall has a distinct advantage for those who plan to take guests or children along because these individuals do not have to pass through the main sleeping area at night in order to use the bathroom, as they would with a rear bath.

Closets and cabinets. You know what those are. Check carefully for door closure fittings and usable space.

Kitchen. You know what this is, too. Again, check carefully for size, capacity, and arrangements for food preparation surfaces and storage.

Swivel chairs. These chairs, usually found in the living room, are mounted on swivel bases to allow them to turn. Most are rocking chairs and some can be moved about. If you decide on these, check your family's seating requirements when traveling. If you should need these chairs for seating while the unit is in motion, look for chairs that are permanently mounted to the floor and are equipped with seat belts. (Because most states prohibit passengers from riding in a trailer when it is in motion, seat belts are not required in these units.) Between the swivel chairs, you will find a table that folds out of the way when not needed.

Tables. We have already covered the dinette and folding table between the chairs. Another type you will run across is not always obvious. This table is found mostly in motor homes. If you examine the floor in front of the couch you may find one or two "carpet plugs"—circular cuts in the carpet. Lifting these plugs reveals holes in the floor. Elsewhere in that unit—in a closet or under a couch—you will find one or two tabletops and two chrome posts that just fit the holes in the floor. Putting the posts in place and securing the tabletops to the posts provides an extra surface in front of the couch for eating, playing cards, or what have you.

Making Your Choices

We have covered the basics of elements that, in some combination or other, go into the floor plan design of just about every RV on the market, both past and present. You should now be familiar with the terminology involved. The next step is to start using it.

Kitchens, some more elaborate than others, will be found in most RVs. Although you do have to think about things like storage and countertop requirements, we can put these aside for now. You *will* have a kitchen!

Permanent beds, dinettes, swivel chairs, couches, and bathrooms are another matter. If, for example, you are leaning toward a fold-down trailer, built-in permanent beds are in the form of slide-out bed ends. We have never seen swivel chairs in a fold-down. However, nothing would surprise us in this creative industry. Dinettes, gauchos, and bathrooms *are* factors. You will face choices like Do I want two dinettes or one along with a gaucho? or Do we really need a bathroom, or Can we get along with a portable toilet? Answers to these questions must be based on your own personal preferences, how much trailer you can afford, and on the size and makeup of your family. Remember, the more elements you add will affect not only cost but also size. Only so many elements can be added to a box before you have to make the box bigger. The next rule that follows is the bigger the box, the heavier it gets—and weight is a factor in gas economy.

Now think about your tow vehicle. How much weight will it haul? If you do not know, then check with the manufacturer of your car, truck, or van. *Do not exceed* the recommended tow capacities! Transmissions and engine replacements get expensive, not to mention the safety factors involved.

Ideally, in the case of trailers, you are always better off first selecting the unit you want to buy, then buying the appropriate tow vehicle. However, do not buy the trailer until you have determined that you can afford the tow vehicle that can pull it. Compromises may be needed. You can see how all of these factors interrelate.

Back to floor plans. Standard-size van campers do not offer a whole lot of flexibility. The choices are bathroom, yes or no, and couch *or* dinette. Wide-body vans, having more space to play with, offer a wider variety of floor plans. Likewise, truck campers offer few choices. It's usually a matter of yes or no to the bathroom question.

Motor homes, travel trailers, and fifth-wheel trailers offer the greatest number of options. Again, if you add elements to the plan, the size of the unit must increase. By the same token, if you have limited the size of the unit, you will have to make compromises: Something will have to go to make room for something else. A dry bath may have to surrender to a wet bath and more closet space. A couch may be replaced by a dinette, or, if you keep the couch, you may have to give up the swivel chairs for the dinette. Remember that except in a fold-down, permanent beds eat up space.

A parting word on size. Many first-time buyers have a fear of buying too large a unit regardless of how they intend to use it. *Forget that fear.* On an extended trip, the more space you have, the happier you will be with your choice. You are used to living in an apartment or a house with a lot more living space than any RV has. You *will* get used to handling any size unit, so do not let fear drive out common sense. The biggest reason people sell or trade their units before they have had reasonable use of them is that they bought the wrong floor plan initially or that they bought a unit that was too small for their needs.

The key word here is *needs.* This is not to say that small doesn't have a place. It's fine for weekends or one- or two-week trips with a small family. But smaller may not always be better *for you.* The hasty trade—or worse, a unit sitting in your driveway because you no longer like it—gets *very* expensive. RVs do depreciate.

Take generous amounts of time in making these decisions and don't be shy about asking questions of your dealer and camper friends. Everybody has questions, even the experienced camper. For the first-timer, they are more critical. Unfortunately, everybody doesn't have this book—but you do!

To help you make your choices, we have included with this chapter a priority form so that you can rate the elements that are important to you. Go to an RV show or local dealer and look over the various options in floor plans and size before completing the form.

Floor Plan Examples

In this section are examples of at least one floor plan for each type of RV. The drawings are not to precise scale, but they are proportional. As you shop and accumulate brochures on the various models in which you are interested, you will discover a wide variety of sizes and floor plans offered by individual manufacturers. The ones shown here are representative of some of the more popular ones.

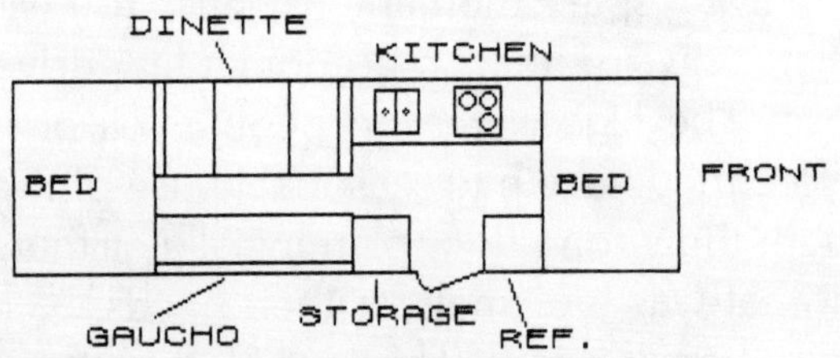

Fold-down travel trailer measuring 22 feet long when open and 16 feet long when closed. (figure 3-2)

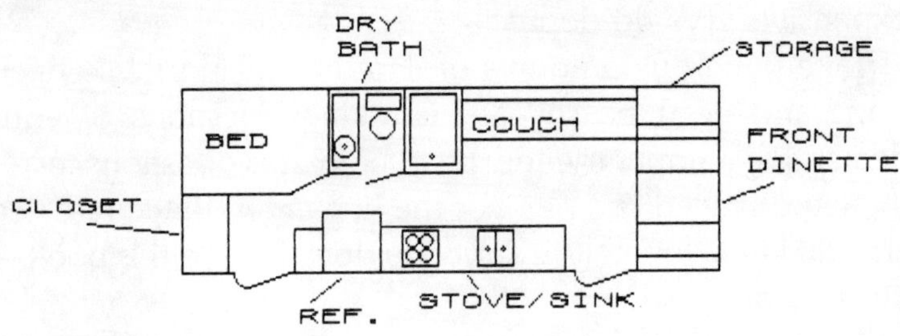

Travel trailer, 27-foot model. (figure 3-3)

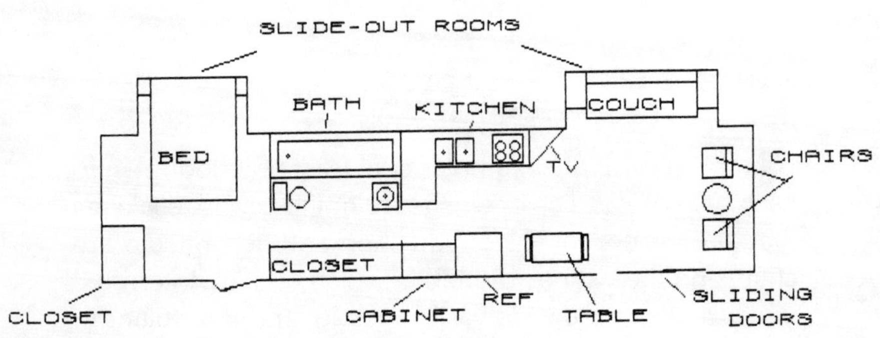

Park model trailer, 35 feet long. (figure 3-4)

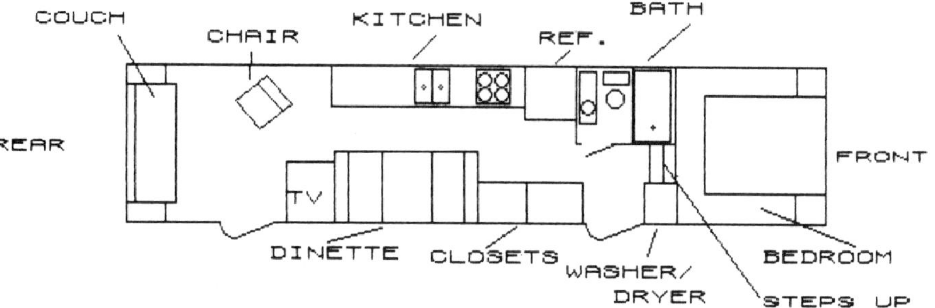

Fifth-wheel trailer, 35 feet long. (figure 3-5)

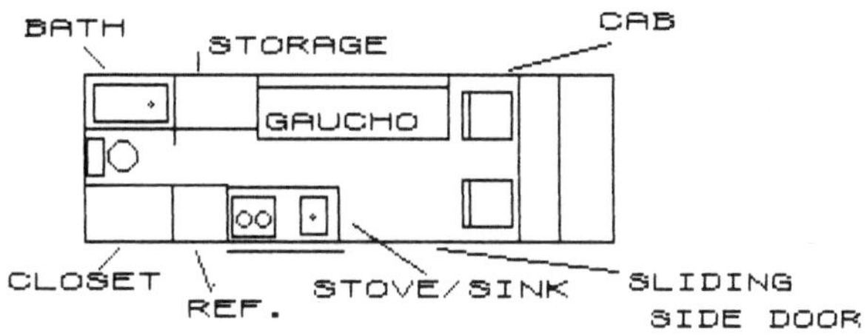

Standard van camper, 19-foot model. (figure 3-6)

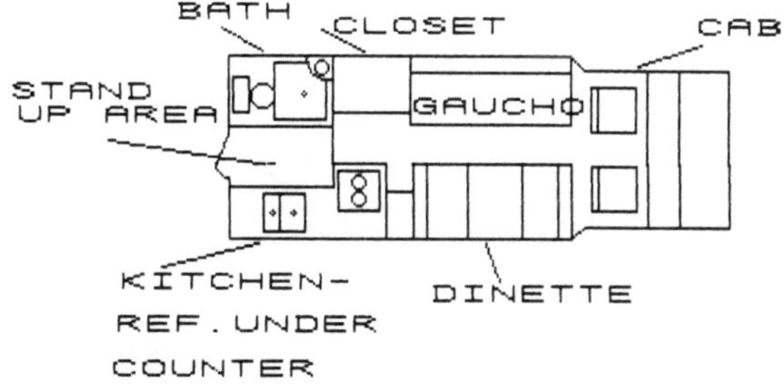

Nineteen-foot wide-body van with wet bath. (figure 3-7)

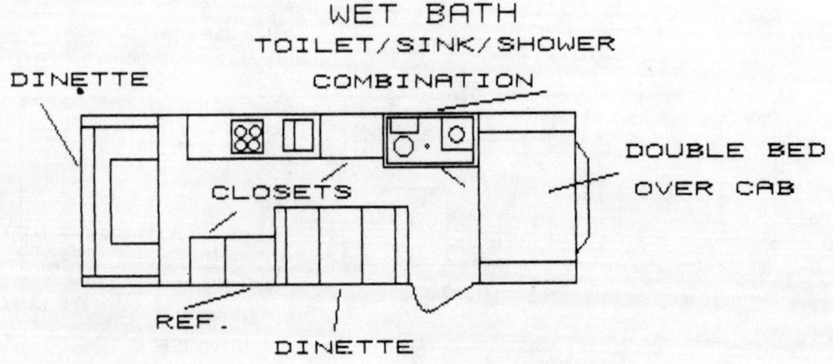

Mini motor home, 23-foot model, with double dinette and wet bath. (figure 3-8)

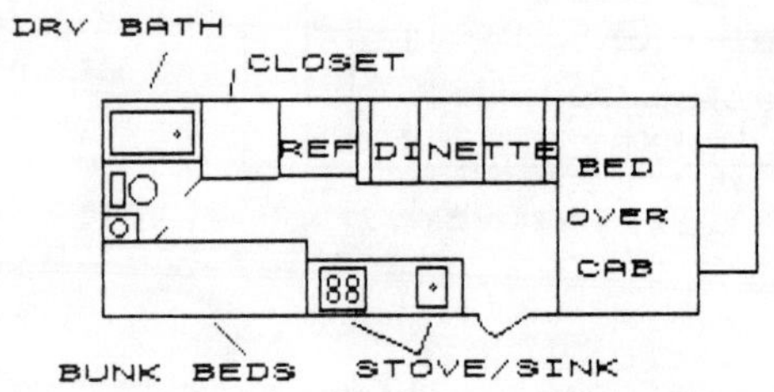

Micromini bunkhouse, 21 feet long. (figure 3-9)

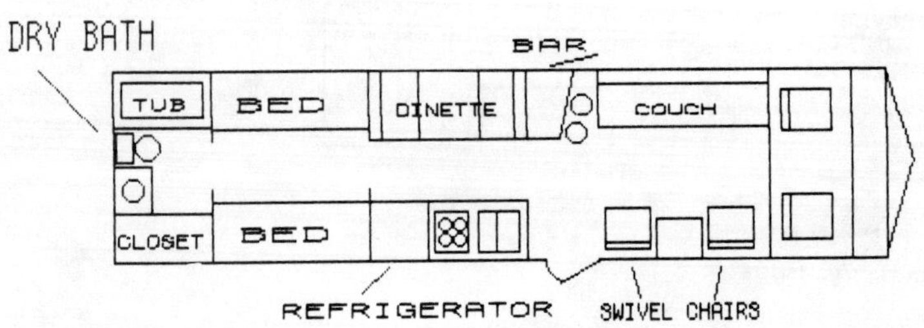

Typical 32-foot class A motor home floor plan. (figure 3-10)

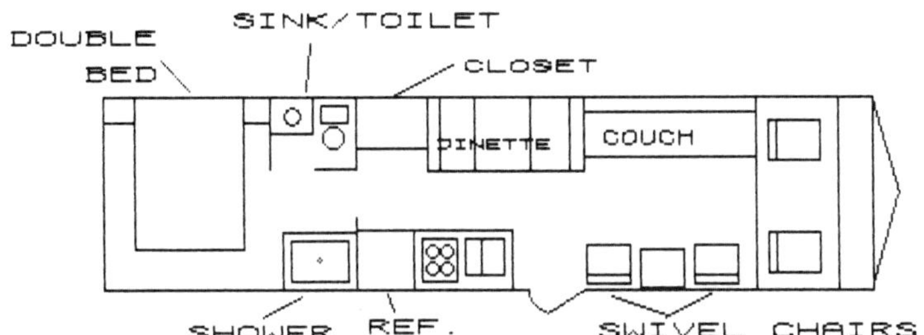

Class A motor home with double bed and split bath. (figure 3-11)

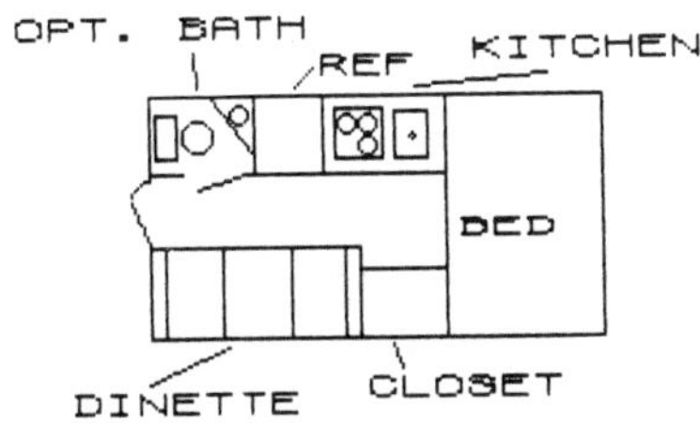

Thirteen-foot truck camper. (figure 3-12)

Floor Plan Priorities

After attending an RV show and having looked at a multitude of floor plans, take some time to rate the importance to you of the various components of a floor plan. On a five-point scale, with 1 meaning what you must have and 5 meaning what you don't want, circle the number that best reflects your current thinking.

	Must have 1	2	3	Don't want 4	5
Gaucho	1	2	3	4	5
Couch	1	2	3	4	5
Dinette	1	2	3	4	5

	Must have 1	2	3	Don't want 4	5
Extra tables	1	2	3	4	5
Swivel chairs	1	2	3	4	5
Permanent bed	1	2	3	4	5
Bunk beds	1	2	3	4	5
Cabinet bunks	1	2	3	4	5
Wet bath	1	2	3	4	5
Dry bath	1	2	3	4	5
Large closet	1	2	3	4	5
Lots of storage	1	2	3	4	5

How many items you rated 1 or 2 will, to a large extent, determine the size unit you should be seeking. Items rated 3 are a compromise. Those that you rated 4 can be given up without too much regret. If you gave an item a 5, forget it.

Do not make more than three compromises in your floor plan. The fourth often is disastrous to your enjoyment of the unit. In the long run, the fewer compromises you make, the happier you will be with your final selection.

Options: Sorting Out Wants and Needs

Chapter 4

Just as in buying an automobile, options on an RV can add to the pleasure of ownership while enhancing creature comforts. And just as in buying an automobile, options can increase the cost of purchase considerably. That is why it is important to learn what some of the more popular options are and what they will do for you in terms of function and what their impact might be on future resale. The following comments are based upon our own personal experience as well as the experience of other buyers.

Air Conditioners

Three basic types of air-conditioning systems are found in RVs: In the language of the trade, they are dashboard air, roof air, and window air. Dash air, which applies only to motorized units, is the same as that in a car. It works only when the engine is running, blowing cool air into the passenger compartment through vents in the dashboard. In an RV, dash air can be expected to cool the driver and passenger seat area but not the entire vehicle. Dash air is a good thing to have when the time comes to sell the vehicle.

Roof air is what the name implies: It is an air conditioner permanently mounted on the roof of the RV. Some fold-down trailers offer this as an option, although the roof has to be braced to support it—and that means a factory installation. Buyers of used fold-downs beware. Do not assume that you can add roof air at some later date, because the roof may not have been braced for it. Most other RVs have been reinforced and prewired for roof air so that it can be added after purchase, but ask if you are not certain. Buyers of used vehicles should watch out for a sagging roof in the area of the air conditioner. It could mean that supports have been weakened by a water leak.

Unlike dash air, roof air requires 110-volt electricity for power, meaning that the cooling unit can be operated only when it is plugged in at a campsite or connected to a generator (the latter will be discussed later). Roof air, which will cool the entire RV, is an excellent choice for both resale purposes and comfort while you own your RV.

Some RV buyers believe that they can save a few dollars by installing in their unit the same kind of air conditioner they have in the bedroom window at home. We do not recommend this practice unless the unit is to be permanently parked on a lot and not towed or driven. RV walls simply are not built to take the added weight, and special bracing is required. In addition to being ugly, these installations can be a safety hazard because they protrude from the side, rear, or front of the unit. Given the two choices, roof air is clearly the way to go.

Generators

These are small engines that produce 110-volt electricity. They are fueled by regular gasoline or propane and can be portable or built into the RV. Built-ins may be purchased as options on motor homes, wide-body vans, fifth-wheel trailers, and—rarely—travel trailers. When installed in motorized units, they usually share the same gasoline tank as the engine but have a shorter fuel draw tube so that they can't consume the last of the gasoline. Trailer installations usually have a separate gasoline tank. Propane-fueled generators are found on motor homes equipped with diesel engines.

Portable generators can be used with any RV; however, their use means that when you park at the campsite you must carry them outdoors from their storage area and replace them prior to departure.

The advantage of the built-ins in motorized units is that the roof air can be operated on the road, thus cooling the entire coach. Generators are always a plus when reselling any unit and are a must in the resale of a class A motor home. They are available with differing levels of power output, so be certain that the one you purchase will handle the total requirements of the appliances you intend to use.

Television Antennas

Many types of television antennas are available for RVs, including roof-rack types for vans, crank-ups for other RVs, and, yes, even portable satellite dish antennas. With the van type, the antenna is incorporated into the roof rack. These are effective, especially if they are equipped with a power signal booster for fringe area reception. The power boost is also recommended for the crank-up type.

The crank-up antenna, which mounts permanently on the roof, can be raised, lowered, and turned for directional reception from inside the RV. Always remember to lower the antenna before driving off. Tree branches can be very effective in removing this option if it is left in its raised position. Also available is a remote-control type that mounts on the roof like the crank-ups but does not raise and lower. This antenna, mounted inside a housing, can be turned for best reception with a remote control from inside.

The satellite dish is great if you must have a choice of many channels no matter where you are, but you pay dearly for the privilege. Some campgrounds now offer cable TV hookups for a fee. To buy this service, you will need a cable receptacle in your unit, but these devices are inexpensive.

Other available antennas include a side-mounting device that attaches to the side of the RV and has a folding antenna inside a tube. These antennas are a bit less convenient to use than the remote or crank-up varieties, but they are cheaper and serve the purpose. Antennas are an asset when reselling.

Microwave Ovens

You are familiar with these appliances, but you may not know that more and more RV buyers are looking for adequate counter space on which to place a small microwave or to accommodate a built-in unit. Powered by a 110-volt source, they are a real benefit for the dedicated microwave cook. Microwave ovens or space for using them can be a big plus at resale time.

Awnings

A couple of basic types of awnings are available: the slide-in and the permanent mount. Slide-in awnings are a bit difficult to erect because the awning must be removed from wherever it is stored and slid into the awning rail that runs along the roofline. More convenient is the permanent mount, which takes one person about five minutes to set up and does not occupy precious storage space. Awnings are a great option because they provide shade for one side of the unit, keeping the interior cooler, and provide shelter from the rain when you want to be outdoors.

Some permanent-mount awnings store in a metal box along the roof line, whereas others roll up to storage position at the roof and have the awning arms attached permanently at the floor line. The box eliminates the attached arms. The choice between the two is a matter of personal preference; both are easy to set up.

Another advantage of having an awning is that you can add a screen enclosure option that converts the shaded area into a screened porch. Equipped with privacy panels, it can become an additional sleeping area for kids or guests. Don't plan to set up the screening if you are only stopping somewhere for a quick lunch, because it takes more setup time than the awning alone.

Keep in mind that rain and wind can bring about the early demise of even the most sturdy awnings. Rain will not harm your awning as long as you remember to lower one of the outside corners to allow the water to run off. High wind is really the most serious threat. Although accessory tie-downs are available for protection against wind damage, the safest bet is to stow the awning when winds are rising.

Electric Step

This is a good option if you are the forgetful type. The electric entry step automatically extends when you open the main entry door and retracts when you close it. A manual step requires that you exit the unit, then grab hold of the metal step and wrestle it into place. Also, if you don't remember to push it back into place before resuming your trip, you may find yourself replacing it at the nearest RV repair shop. High curbs, stumps, and rocks are lethal for forgotten extended steps. On the other hand, if you forget the electric step the first time you enter the RV from the outside and bark your shin as it extends, *do not* remember that we said it was a good idea.

Other Small Amenities and Conveniences

Cruise control. Very important for resale. It works just like the one on your car. Get it!

Monitor panel. This is a handy small device that takes up little wall space inside the unit and provides such necessary information as how much fresh water you have left in the on-board tank, how full the wastewater holding tanks are, and the condition of your batteries.

Water filter. If you are concerned about the quality of your drinking water wherever you go, one of these devices can be a good thing to have. The type that attaches to the water supply pipe under the sink is probably the best.

Emergency start switch. Another good thing to have. This is primarily a motor home option, although we have seen trailers and tow vehicles similarly equipped.

Some backgrounding is necessary in order to explain why an emergency start switch is a desirable option. RVs have two types of batteries: one to start the engine and the other to provide power for the living area of the unit. The engine battery is just like the one in a car. The living area battery is known as a deep-cycle or marine battery. This power source operates the interior lights, water pump, and furnace blower when the RV is not connected to 110 volts at a campsite. The living area battery differs from the engine battery in that the deep cycle will take more repeated charging and discharging. The higher the ampere-hour rating of the battery, the longer it will last between recharges.

A device called a battery eliminator usually is installed between the two batteries to prevent the deep cycle from draining the engine battery but allow both batteries to charge when the engine is running. The emergency start switch is a button installed on the dashboard that allows the driver to bring the deep-cycle battery into the starting circuit if the engine battery happens to be low. This device is worth the extra money it costs. When you are miles away from a repair facility, you'll wish you had one if you don't.

Roof rack and ladder. Usually a motor home option. Provides for easy access to the roof and for extra storage.

Spare tire. Obvious. And yes, they are optional.

Rustproofing. Buy it. This applies to vans, minis, microminis, and tow vehicles.

Most other options are such frills as pushbutton bars, built-in countertop blenders, and remote control spotlights. Saying yes or no to such options depends on personal tastes and needs. A trip to the camping equipment store or your dealer will enlighten you as to all of the many possibilities. We believe that we have covered the most important options that could affect your comfort and the ultimate resale of your RV.

A few other items now need discussion: the refrigerator, furnace, converter, and the engine in motorized RVs and tow vehicles.

Refrigerators

This appliance is standard on some RVs and optional on others. When optional, the choice is between a refrigerator and an icebox. When standard, the choice is usually size. RV refrigerators normally operate on 12-volt battery, 110-volt electricity, and/or propane gas. Most RV refrigerators will operate in at least two of the above modes. For example, one type will run on 12-volt battery or 110-volt electricity, using a converter, and is popular in vans and fold-down trailers.

Another two-way type, operating on 110-volt electricity and propane, is common in newer RVs having larger refrigerators. They operate on propane when traveling and will not run from the battery. Because all pilot lights must be extinguished when refueling, it means restarting the refrigerator afterward—a minor inconvenience. Still another type will function in any one of three modes. However, veteran campers seem to prefer propane operation because when stopping to sightsee or shop, they don't have to worry about finding a run-down battery and a warm refrigerator when they return. If a large-size refrigerator operates on battery without the engine running, the battery most likely will be dead within 30 minutes or so. In this case, the only solution is to remember to switch to propane before leaving the RV. A battery charge seems to last longer with the smaller 12-volt refrigerators.

A new generation of larger refrigerators provides three-way performance, but these appliances are equipped with an automatic changeover. They will automatically seek the optimum operating mode without your having to remember to manually change it.

Because of the way an RV refrigerator is designed, it must be leveled if it is to stand still for extended periods or it will cease to function. If it is not leveled in this situation, the refrigerator's coils can plug up, resulting in a repair bill that will not soon be forgotten. As long as the vehicle is moving, the rocking motion is enough to keep the coils from corking up. The exception is a new type of refrigerator that eliminates the need to level the RV when it is parked for extended periods. Because they are new to the market, these refrigerators probably will be scarce for a while.

Get the largest refrigerator that is available for whatever RV you buy. If you are buying a used vehicle, you won't have a choice, but at some future time you can always upgrade if you decide to do so. However, the price of a larger unit may make you decide to make do with the one you have. They cost far more per cubic foot than home refrigerators.

Leveling the RV

So how do you level an RV? It's easier than it sounds and takes less time than it takes to read about how to do it. First, you head for a camping store and buy two self-stick bubble levels. Then, with a friend, find the most level spot to park the RV that you can. Shopping mall parking lots are usually a good bet if your own driveway isn't suitable. Place the two levels on the bottom of the freezing compartment of the refrigerator. Arrange them so that they are at a 90-degree angle to each other. The object here is to level the refrigerator front-to-back and side-to-side. Now, with your friend watching the levels, move the RV slowly around to a spot where both levels show the bubble in the center. Your refrigerator is now level. Remove the levels from the freezer and take off the protective backing. Center the bubble and stick one to the outside on the *side* of your RV and the other on the *front.* Trailer owners usually place one level on the right front corner (passenger side) on the front and the other on the right front corner of the side (see figure 4-1). That way you can see both levels at the same time. Some motor home owners prefer to mount the levels inside, with one on the dash and the other on the driver's side door—or if there is no door on that side, on a side panel.

Wherever they are placed, whenever both bubbles are centered on both levels you can be assured that the refrigerator is perfectly level. Fortunately, most campsites are fairly level. Frequently, by maneuvering the RV around the campsite, you will find a spot where it will be perfectly level. If not, more action on your part will be needed. An RV can be leveled in several ways. Choices are limited only by the amount of money and effort you wish to spend to accomplish it. The least expensive method involves making (or having made) ramps from 2-by-6-inch boards (see figure 4-2). You will need at least two sets if you own anything other than a fold-down trailer. Fold-downs usually come equipped with built-in stabilizing jacks for minor leveling chores, but if yours does not have them, you will need one of the sets in figure 4.2.

Use the ramps stacked, as shown in the illustration, or singly. Some trial and error is necessary to find the best combination of placing boards under the wheels to level the unit. Trailer owners have it a little easier because they need level only side-to-side with the ramps (the tongue jack will take care of front-to-back leveling).

Other methods of leveling involve varying degrees of expense. Discuss the options with an RV dealer and select the one that is best for you. The options include ramps made of various materials, built-in scissor jacks, and power leveling jacks. The last are the easiest to use and, naturally,

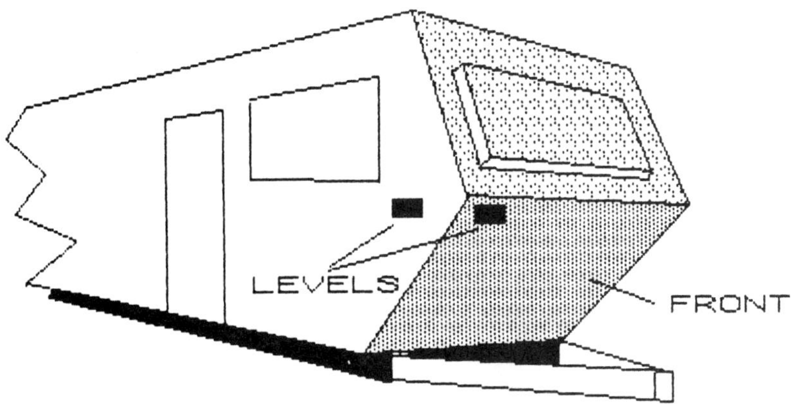

Placement of levels on a travel trailer. (figure 4-1)

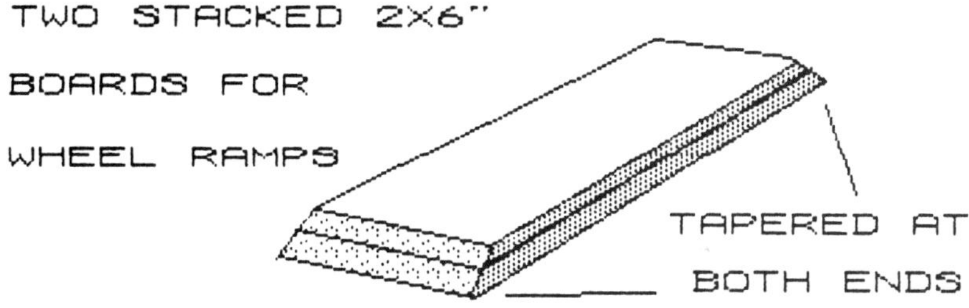

Two stacked 2- by 6-inch boards to be used as wheel ramps in leveling an RV. (figure 4-2)

the most expensive. Whatever method you choose, make sure that the refrigerator is *level.*

Furnaces

An RV furnace operates very much like the home variety but is smaller. It has a heat chamber fired by propane and has a 12-volt blower plus a thermostat for temperature control. Standard equipment on larger RVs, furnaces are optional in many smaller units. When there is a choice, opt for the biggest furnace available for your unit unless you never plan to travel when it's cold outside.

Older RVs were sometimes equipped with simple heaters that did not have blowers or ducts to distribute the heat, possibly resulting in cold areas. If you are buying a used vehicle, you will want to find out what kind of heating system it has.

Converters

On most newer RVs, converters are standard equipment. A converter is an electrical box that converts electricity from 110 volts to 12 volts. Because most of the equipment on an RV is designed for 12-volt operation, except for the roof air conditioner, the microwave oven, and equipment plugged into the 110-volt wall outlets, it is necessary, when plugged into a 110-volt campsite line, to convert that power to 12 volts to avoid burning out the wiring and the 12-volt appliances. That is what converters are for. Usually, they are equipped with a battery charger that will charge the deep-cycle battery when plugged into 110 volts.

Again, older RVs may not have converters. Instead, they were fitted with two-way light fixtures consisting of two separate sockets with separate switches, one for 110 volts and the other for 12 volts. If you spot this type of fixture while shopping for a used RV, suspect the absence of a power converter. This doesn't mean the unit should be shunned, however; it's just that that's the way things were done in the past.

Engines

If there is a choice of engine sizes for a tow vehicle, buy the biggest one available. You won't be sorry if you have to drive in hilly or mountainous terrain. With motor homes, there is usually little choice; however, their engines are more than adequate for anything short of four-wheel drive country. The exceptions are the four-cylinder engines in older microminis and the diesel engines in lower priced motor homes. Unless you are strongly biased in favor of the diesel, think twice about buying an RV so equipped. In some parts of the country, you will take a terrible beating at resale time—if you can sell it at all. More on engines in chapter 6.

Tow vehicle buyers should be certain to buy a complete tow package that includes an oversize radiator, if available; tow mirrors; a transmission cooler; a heavy-duty differential; a brake control, if the trailer has electric brakes; a larger alternator, if available; a hitch that is adequate to the task, with a margin for overload; wiring for brakes and lights; and sway control, if a travel trailer is to be towed. Some of these items are available from the tow vehicle dealer and others from an RV dealer or hitch specialist. Don't skimp on these features, because your safety is involved.

New or Used?

Chapter 5

A major factor in any purchase decision is how much you can or want to spend. By now you should have a pretty good idea of the type of unit you prefer and the floor plan and size you like best. The issue now becomes one of whether you are going to go for a brand-new model or if you will settle for a used unit in good condition. As with everything else, there are advantages and disadvantages to either choice. Let's take a look first at the buy-new alternative:

Advantages in Buying New

- You can order the interior color of your choice from the selection offered by the dealer.
- You can order factory-installed options in desired combinations.
- You will get a full warranty on the chassis, appliances, and options. These warranties vary from manufacturer to manufacturer, so compare carefully and read the fine print. With motor homes, it is more common to

take the coach to the dealer of the chassis manufacturer for warranty repairs to the engine and drive train and to the motor home dealer for repairs of appliances and options.

- You can maintain the unit right from the start according to your own standards.

Disadvantages in Buying New

- Depreciation! Like automobiles, RVs depreciate more rapidly in the first three to four years of ownership, with the biggest drop occurring in the first two years. Fortunately, RVs generally depreciate at a lower rate than cars. You make out best if you keep a unit for several years before selling or trading it.
- Expect to spend some time at the dealer's for repairs. No matter how well a dealer prepares an RV for delivery, things can go wrong. We have yet to find a perfect RV at any price. If you have ever purchased a new home, you know what we're talking about. In this case, you are buying a *home on wheels.* Fortunately, a full warranty is backing you.
- The initial investment is usually much higher than for a comparable used unit.

Advantages in Buying Used

- Lower initial investment. The previous owner takes the depreciation lumps.
- Sometimes you can wind up with more options than you would have buying new and get them at a much lower price. Options depreciate in value, too.
- By careful shopping, you can buy a unit of much higher quality than you could afford to buy new.

- The first owner probably had any major defects corrected while the warranty was in effect. But don't count on it.

Disadvantages in Buying Used

- You get a limited warranty or no warranty at all.
- Service records are lacking. You are more likely to run into this on an RV that has had two or more owners, with the result that you don't know how well the unit has been maintained.
- There is no interior color decor selection and no selection of options.
- Problems are more likely to develop because of age and use. The older the unit, the greater the risk—but that is why the initial investment is lower.

A buyer should always consider used RVs along with new ones, with personal circumstances dictating the way to go. With patience and perseverance, some good deals on used units can be found.

Tips on Buying Used Units

If you have decided to try a used RV, whether a trailer or motor home, here are some tips:

1. Shop new first. Get some firm "buy" prices on the unit in which you are the most interested. This will give you price comparisons when shopping for a used vehicle. Be certain to compare apples with apples in terms of equipment. Look at overall condition. Does the unit look as if it has been abused?

Beware of private individuals trying to sell you a travel

trailer or park model—new or "like new"—at an unbelievably low price. These trailers usually carry an offbeat brand name that changes with every other unit coming off the line. They are built to last about as long as their last-used brand name. Individuals have been known to approach people in campgrounds or through newspaper ads attempting to sell this "like new" trailer for "a fraction of what it is worth." These approaches are usually coupled with a sad story of a death in the family or a desperate need for money. Once a sale is made, they head back to the factory for another one and then look for another campground and their next customer. Don't let it be you! There are legitimate people out there who have true tales of woe and are attempting to sell RVs. Just watch out for weird brand names and cheap construction coupled with "the once-in-a-lifetime price."

2. We would turn down anything shaped like a shoe box, where *no* attention has been given to aerodynamics. Most of these units were built in the 1970s or before. This is not to say they were not functional, because most were; however, budget can often overrule beauty. You will note that some manufacturers put some curves or bends on the front ends to give their RVs some character. We would choose these over the plain box.

3. We run the other way when we see anything with more than a perimeter seam on an aluminum roof exterior. For example, some mini motor homes built during the 1970s had a roof that looked rather like a trapezoid when the vehicle was viewed from the side. Granted, it was good looking, but how many seams must it have taken to form that roof? It looked like a road map. One owner we know spent many an hour trying in vain to stop the roof leaks—at least three times a summer for all the time he had it. The leaks persisted, and eventually all of the roof supports had to be replaced. A fix is now possible in the

form of a rubberlike sheeting that can be applied over the old seamed roof. We have not priced this process, but we suspect that it is expensive.

4. Check the condition of the roof exterior. If roofing tar—it's black and thick—has been applied to the seams, move on to another RV. If, on the other hand, the seams are covered with a white or silver substance, it is probably a special RV roof sealant. One of two things happened: Either the previous owner was doing preventive maintenance or there was a roof leak. In either event, the presence of the sealant demonstrates some care and concern.

5. Check the interior for roof and window leaks. Open all upper cabinet doors and look for stains or warping of the ceiling. If you spot stains, push up in the spot to see if it is soft. If it is soft, there could be serious damage to the roof joists. This is an expensive repair job unless you can do it yourself. All RVs can leak, even new ones. The key is in how quickly the leaks are caught and repaired. Let your nose help you here, too. If you detect a damp, musty odor when entering the unit, suspect leakage problems.

6. We would avoid anything that has a carpeted or foam-covered *ceiling*! Carpeting in that location can trap cooking odors, is hard to clean, and covers a variety of leakage problems. The type of foam to which we refer looks like the surface of a gravel road; in addition to a tendency to chip off, it shares the disadvantages of carpeting. This problem is not easily fixed.

7. Soft floors can also be a problem. If a leaky water tank was not repaired right away, the floor joists underneath could have been softened—another very expensive repair job. So move around the unit and be aware of any spongy areas in the floor. Check for water damage to the floors of the outside storage compartments and also to the toilet and sink areas.

8. Anything with shag carpeting on the floor (a disaster to clean) is a turnoff. You will want to replace the carpet. Although removing and replacing carpet is a major project, in most cases it is not very expensive.

9. Check upholstery for evidence of undue wear or sun rot.

10. Inspect drawer fronts and cabinet doors. During the 1970s, some manufacturers decided to use plastic for cabinet doors and drawer fronts. We don't have to tell you the result. You will find more than one instance of cracking! Newer plastics used in more recent models are tougher and seem to hold up well.

11. Inspect the exterior carefully, watching for newly painted or replaced body panels. Severe damage could have bent the frame. Also, look for pitting. Some units built in the 1970s having a bonded aluminum skin on the exterior had a most undesirable propensity to pit. With some, the pitting seemed to settle mostly around the wheel wells, under doorways, and directly below windows. With others, it appeared to be more widespread. Part of this, we suspect, may have been due to owner neglect of water leaks. The pitting does not seem to affect the units structurally unless there has been extensive water damage within the walls, but it is unsightly. In later models, an improved bonding process eliminated the problem. In addition, most companies later switched to fiberglass skin, which does not pit. We would move on to another unit if pitting is present.

12. If, after having made preliminary careful checks, you are still interested in the unit, put down a small deposit on it, assuming, of course, that you have negotiated what you consider to be a reasonable price. Be sure you have it in writing that the deposit is refundable and, in the case of a motorized unit, that the sale is subject to the approval of your mechanic and/or passing a test drive with all lights,

systems, and appliances working. If you are negotiating with a dealer, ask if an extended service agreement is available. If it is, definitely add it to the terms. These are insurance policies that will help to protect you in case of mechanical or appliance failure within prescribed mileage and/or time period. These policies usually have some form of deductible.

13. Check the "book" value of the unit you are considering. Like many automobile dealers, RV dealers do not normally share this information with the public, so you will probably have to get it from a bank or insurance company. Book values usually are based upon average condition and do not include options or factors for high or low mileage. The books contain figures for these variations that have to be calculated separately. Don't be afraid to pay more than "book" for an exceptionally clean and desirable unit. Consider instead what a new one would cost.

14. With a motorized unit, have your mechanic test drive it and give the mechanical areas a thorough examination. Usually, you will have to take the mechanic to the unit rather than the other way around. Most private parties or dealers will not let you remove the RV unless a staff member or the owner goes along, and sometimes this is not possible. It is worth the added expense to have your mechanic check the vehicle, especially if no warranty is involved. Before involving any mechanic, test drive the RV yourself. You may decide that you do not like something about it and so can save the expense of hiring a mechanic.

15. We would watch out for mid '80s and earlier four-cylinder engines on micromini motor homes or wide-body vans. Many were underpowered for the amount of vehicle they were being asked to haul around. This is especially true of any four-cylinder diesels you may run across.

16. Ask for a complete demonstration of all options and appliances to make certain that they are in good

working order. Normally, this is done at the time you take delivery. Whether you are dealing with a dealer or with a private party, make sure that you have a written guarantee that everything will work at the time of delivery. Pay special attention to the refrigerator. An RV refrigerator takes from six to eight hours to get cold, so make certain that it is turned on, using *110-volt* current, the night before your demonstration appointment. If it works well using electricity and if it is equipped with a propane option, ask the seller to light the pilot light. If it lights, the refrigerator is probably in good shape. (See point 19.)

17. Other options and their quirks to which you should pay attention include:

Roof air-conditioning.

Obviously, it should deliver cold air. However, some units have a time delay, perhaps as long as five minutes, before the compressor starts. If it takes longer than this, or if the compressor doesn't start at all, something is wrong. If the fan does not start when the unit is turned on, suspect one or more of the following reasons:

- The circuit breaker may be off.
- The RV may not be connected to an outside power source.
- If the RV has a microwave oven, there may be a switch that selects microwave or air-conditioning. The switch is there to prevent both options from being operated at the same time because of the large power drain.
- If the RV is generator equipped, the air conditioner may be connected to the generator instead of to the outside power source. The outlets for both generator and outside source are usually located in the generator compartment. Switch the plug to the other outlet and proceed with your test.

Generator.

The generator should start easily and smooth out after a few seconds of running. Normally, it slows down when a 110-volt appliance is turned on, but it should *not* stop. Have your mechanic check this option.

Furnace.

Like any propane-operated appliance, the furnace will light more easily if the air is bled from the propane line first. Do this by lighting one burner of the stove and letting it burn for a few minutes. Then light the furnace. If it won't light, have the RV dealer check it. (See point 19.)

Power converter.

Try the interior lights first on battery, then plug the RV into an outside power source. The lights should get brighter. If they do not, suspect a problem; check for a blown fuse or tripped breaker on the converter, which may have a battery/AC switch. Make certain it is in the proper position. Failing that, have a dealer check it for you.

18. Have the water system demonstrated. Look for leaks and, if the unit is so equipped, check that the hot water system works. Specifically check:

Water pump.

The pump switch, usually found in the kitchen area, is either a separate switch or is incorporated into the monitor panel. With water in the system, turn the switch to the "on" position. All faucets should be turned off. You should hear the pump turn on, and then, as pressure builds in the lines, it should shut itself off. Go to a faucet and turn it on.

Use caution, because air in the lines may cause some initial spurting. When the water is running smoothly, turn the faucet off; the pump should also turn off. This is called a demand water system; that is, the pump runs only when water is being used. If it continues to run with all faucets off, suspect that there is a leak in the water system or supply pipe, that the water tank drain is open, or that the demand switch on the pump is defective. The last possibility would need to be checked by the RV dealer if you have ruled out leaks or open line drains. The switches are inexpensive to repair. If the pump fails to turn on at all, check for a blown fuse.

Hot water heater.

The operation is similar to that of a home unit. Make certain that there is water in the system. Most heaters have lighting instructions on the panel of the heater. Remember to bleed the air in the propane lines by lighting the stove. Once the pilot is lit, the main burner should cycle on and off as the water reaches the selected temperature. If the pilot should fail to light, there may be dirt or cobwebs in the pilot tube. Removing the tube and blowing compressed air through it usually corrects this problem. If not, the tube is simple to replace. Some newer models are equipped with electronic ignition, so there will be no pilot light. Check for burner operation. (See point 19 below.)

19. Check for propane leaks. *Do not use a match.* Use your nose. If the RV has been closed up and you notice a strange odor when you enter, suspect a propane leak. *Do not perform the above equipment tests until the problem is resolved.* We recommend that the RV dealer be asked to test for leaks. Dealers have the proper equipment to do the job.

20. Ask if the operating manuals are available; if they are not, ask the dealer to obtain them. Do not expect miracles. They may be unavailable from the manufacturer.

21. High mileage on a motor home does not always mean that it is worn out mechanically. Remember that motor home mileage is mostly highway miles (vans may be an exception), which are easiest on engine and drive train. In contrast, a passenger car typically is driven through stop-and-start traffic, which is the hardest on the mechanical components.

Tips on Buying New *or* Used Units

Some additional tips for buyers of either new or used units:

1. A one-piece roof is better than a seamed roof because there are fewer places for it to leak. Another good bet is the one-piece roof that rolls over the side, eliminating the perimeter seam, but such roofs are rare. Rubber-coated roofs, new to the market, sound like a good idea. Only time will tell. Usually, in the case of aluminum, a one-piece roof is better than a seamed roof.

2. Pay attention to wall construction. If you are looking at new units, ask the dealer to show you a cross section or drawing of the wall construction. If you see a lot of staples holding everything together, you might want to consider another manufacturer. Remember, in most cases this unit is going to be traveling down the road bouncing and flexing. You don't want to turn around and find your upper cabinets occupying your couch. In used RVs, look for any buckling on the outside skin, especially around doorways. It could mean that the body is bending due to inadequate construction or from abuse on very rough roads.

3. Compare the individual merits of aluminum and fi-

berglass exteriors. Both have their good and bad points. Aluminum will hold up indefinitely if cared for, but so will fiberglass. Both should be kept waxed. (On fiberglass, use a fiberglass wax—*not* a car wax.) Aluminum will bend or dent if struck, whereas fiberglass will crack. Cracks in fiberglass can usually be filled in and repaired, but when aluminum is damaged, it is sometimes better to replace the skin. In most cases, fiberglass is more expensive to insure, but we believe that it is easier to keep looking good.

That's about all we can tell you about construction without getting into specific makes of RVs and how each is made, something that is not within the scope of this book. Suffice it to say that some RVs are built better than others. As you shop and ask for details of construction, the well-built models will quickly become apparent. Construction is a major factor in repair and maintenance cost, not to mention the matter of being unable to use your vehicle while it is being repaired.

4. *Gross vehicle weight rating* (GVWR) is a term that often confuses the buyer. Somewhere on the RV, a sticker or embossed plate gives the GVWR. This is the *maximum* (loaded) safe weight the RV is designed to carry. It is *not* the dry weight (empty of water, propane, food, people, etc.), which is usually given in the manufacturer's brochure. When selecting a tow vehicle and calculating its hauling capacity, use a weight factor that allows for the GVWR of the trailer.

5. Cost is another major factor in the used versus new decision. Specific guidelines are elusive. Costs vary by region, condition of the unit, quality, age, and equipment. If you shop for a new model, look for highly discounted units or units that are new but are the previous year's models. Good deals on new *do* exist. Remember, too, that although trailers at first appear to be far less expensive than motorized units, you also need an appropriate tow vehicle

to pull one. Add the two together and the cost comes out about the same as that for a motor home.

6. As suggested earlier, before buying, check the "book" price of the specific used RV in which you are interested. Sometimes new RVs are also listed in the books; if they are, the prices given will reflect the manufacturers' suggested list prices. Because they do not always reflect true market value, book prices should be used only as guides. Demand for a particular model that is in short supply can drive a price well above the one listed in the book.

Use the checklist in appendix A if you are buying a used RV. If you comparison shop for best price, it is doubtful that you will pay the sticker amount for either a new or used vehicle. So shop, shop, shop!

Getting Down to Choices

Chapter 6

Anyone interested in buying an RV must face the fact that there are hundreds of choices. If you pick up an RV magazine at a newsstand, there will seem to be more RV manufacturers and brand names than there are trees in the forest you visualize as a future campground. But don't panic: The number of brands available in and around where you live will not begin to represent the total number available. This is because certain manufacturers have more market penetration in some parts of the country than in others. Unless you are willing to travel some distance to purchase a specific brand, your choices will be limited to those brands carried by your nearby dealers. On the other hand, some people will travel hundreds of miles to buy what they are convinced is the perfect unit, even though they know they will have to travel those same miles for warranty service.

Engines and Chassis

For the motor home buyer, more good news. You will *not* be faced with many engine or chassis choices. A shopper for a new class A motor home will find that most units are built on a Ford or Chevrolet truck chassis. The Ford will carry a 460-CID (cubic-inch displacement) engine and the Chevy a 454 CID. John Deere has recently entered a contender in the chassis arena, so you may run across this one as well. The engines used have been on the market for years and are built for the long haul. Finding replacement parts should not be a problem.

A shopper for a used class A will run across many older units that were built on Dodge chassis and equipped with 318-, 360-, 413- or 440-CID engines. These are good engines, although in our experience they had a tendency to flood rather easily while being started. At this writing, Dodge no longer makes a class A chassis; however, obtaining parts for existing motor homes has not been a big problem. A great many of these engines are still around in the older motor homes—which says something positive about their durability.

Shoppers for new or used mini motor homes will discover the same engines under the hoods of the van chassis, although the 350-CID Chevy is used rather than the 454. A 454 on a minichassis is rare indeed as also is the 440 on a used Dodge minichassis. A word of caution about the Chevy 350-CID engine. If it burns gasoline, fine, but if it is a diesel, beware. This engine was built in the late 1970s and early 1980s when, during a surge in popularity for diesels, Chevrolet converted some 350-CID V-8 gasoline engines to diesel. Another engine to consider avoiding is the Renault built diesel found in certain domestic, wide body, van type motor homes built in the early to mid

1980s. Our customers reported lots of problems with these. If you buy diesel, buy an engine that was *designed* to burn diesel fuel.

In ordering a new motor home, you will find that if a particular brand is built on a particular brand of chassis, it is unlikely that you can choose a different chassis. There are exceptions, of course.

While shopping, you may also run across another peculiarity of the RV industry. You may find a 1988 motor home (or any other year's model) built on a chassis that is one or two years older than the year the body was built. In other words, a 1987 chassis with a 1988 body. No, they are not trying to hoodwink you. It is not a used chassis with a new body. Both are new. In this example, the motor home is considered a 1988 model for purposes of valuation. The model year of the body is the year of the motor home.

Manufacturers buy chassis in quantity to get a better price, but if they buy 300 chassis in 1987 but build only 250 motor homes that year, they will use the leftovers the following year. The advantage to you is a possible savings of a 1988 chassis price hike—if you buy early in the season. This also explains why a buyer may not have a choice of chassis with a particular make or model of motor home.

Among more expensive motor homes, you will encounter chassis manufacturers different from those mentioned above. Some will be bus chassis or a variation; some will have rear engines; and some will have large bus or truck diesel engines. We are not aware of any major problems with these.

Vans and microminis are also built on chassis that, for the most part, have good track records in terms of durability. Another interesting thing to know is that RV manufacturers may modify, or order modification to, the suspension system. As a result, the ride may be different from brand to brand, even though the chassis are the same make. The test drive will point up any differences.

Body Construction

In our experience, most RV manufacturers have always tried to construct a body that will hold up well to the abuses of the open road. With some exceptions that were mentioned in chapter five, failure of units to measure up is frequently due to owner abuse or neglect. Usually, the exceptions involved a new material or manufacturing technique that did not prove unsatisfactory until the units were in use. Such failures are rare on newer RVs because most manufacturers have learned from these experiences and have corrected the problems. The number of 20- and 30-year-old RVs still in use will attest to the manufacturers' desire to build a good product and their success in doing so. You will see many of those older units as your camping experience grows.

As noted in the previous chapter, it is not within the scope of this book to evaluate the many different construction techniques. Question dealers thoroughly on this issue.

You will find variations in wall, floor, and roof construction, with some manufacturers using all-wood framing, a combination of wood and steel, all steel, all aluminum, or other combinations. Because of the wide variety of construction techniques represented in the market, be sure to find out which were used in the units you look at. Remember what we noted previously about the use of staples to hold things together. Almost all manufacturers use staples somewhere in the unit, but if staples are the only things holding a *wall* together, you could be buying trouble. To repeat: Because you can't see through walls, ask to see a cross section of the wall construction and ask about the use of staples. Also remember that if you are alert to water leaks and take care of them promptly, and if you don't abuse your unit, you can look forward to many happy years of camping with just about any RV you buy, with exceptions as noted above. If you are buying a used unit, bad construction will be apparent even to the untrained eye. Heed the warnings in the previous chapter.

Insulation is another factor to consider when buying new (you won't have a choice in a used RV). The higher the "R" factor, the better the insulation, so ask your dealer.

Appliances

Choice of brands of appliances are not in the cards, for the same reason as chassis—the manufacturer is likely to buy in quantity from one supplier. Happily, the brands are recognized and respected in the industry—Coleman, Dometic, Duo Therm, Magic Chef, Norcold, and Siber, to name a few. All will do what you need done. If certain features of a particular appliance brand are important to you, it will probably mean a switch in make of RV to get them. When buying used, of course, you will not have a choice. However, if you replace an appliance, the world is your oyster. Parts are readily available, in most cases even for older units.

Trade Names and Price Ranges

The section that follows gives the names and price ranges of some new RVs that, after much discussion, we decided we would consider buying for ourselves. This is not to say that *you* should not consider others. You should! Shop and compare. The list is based on two factors, the first of which is our opinions of the condition and durability of the hundreds of used RVs seen at close range in our experience. We believe that the condition of a used RV can be valuable when evaluating a new one, especially if the used model has suffered some abuse and weathered it well. The second factor in building the list was the inclusion of some units that have been around a while and appear to have a good track record. Our list is *not* all-inclusive; it is meant to be used as a starting point.

Price range categories are approximate, with some overlap of ranges, up or down, in some of the makes listed. If the overlap is large, the brand name may be listed in more than one category.

Again, the list is intended as a rough guide to help you search out makes and models that will fit your budget. Ranges given are base prices for new RVs. Brand names and/or manufacturers are listed alphabetically. Even though particular trade names have been listed, you should still check construction details, because specifications can change.

Trailers

Fold-Downs
(*Range:* $2,000 to $5,000)

Coachmen
Coleman
Jayco
Palomino
Starcraft
Viking

Travel Trailers

(*Range:* $4,000 to $15,000)

Aluma-Lite
Camino
Coachmen
Hi-Lo
Jayco
Layton
Mallard
Nomad
Play-Mor
Prowler
Shasta
Starcraft
Sunline
Terry
Wilderness

(*Range:* $15,000 to $25,000)

Argosy
Carriage
Ceville
Classic
Country Club
Designer
Presidential

(*Range:* $25,000+)

Airstream
Avion
Imperial
London-Aire
Royals International

Park Models

(*Range:* $12,000 to $17,000)

Layton
Nomad
Play-Mor
Yellowstone

Fifth-Wheel Trailers
(*Range:* $8,000 to $20,000)

Aluma-Lite
Coachmen
Layton
Mallard
Nomad
Play-Mor
Prowler
Shasta
Starcraft
Sunline
Terry
Wilderness
Yellowstone

(*Range:* $20,000+)

Carriage
Holiday Rambler
Royals International
Yellowstone

Motor Homes

Van Campers (Standard)
(*Range:* $18,000 to $30,000)

Falcon
Fiesta
Horizon
Xplorer

Van Campers (Wide body)
(*Range:* $25,000 to $45,000)

Trans Van
Travelcraft
Turtle Top

Mini Motor Homes
(*Range:* \$20,000 to \$35,000)

Coachmen
Itasca
Jamboree
Mallard
Shasta
Tioga
Travelcraft
Winnebago
Yellowstone

(*Range:* \$35,000+)

Aluma-Lite
Born Free
Holiday Rambler
Jayco

Micromini Motor Homes
(*Range:* \$20,000 to \$26,000)

Coachmen
Itasca
Shasta
Sunrader
Winnebago

Class A Motor Homes
(*Range:* \$30,000 to \$55,000)

Allegro
Aluma-Lite
Bounder
Coachmen
Cross Country
Cruise Air
Itasca
Mallard
Pace Arrow
Shasta
Southwind
Travelcraft
Winnebago

(*Range:* $55,000 to $100,000)

Airstream
Barth
Coachmen
Cross Country
Foretravel
Holiday Rambler
Revcon
Travelcraft
Vogue
Xplorer

(*Range:* $100,000+)

Barth
Beach-Craft
Beaver
Blue Bird Wanderlodge
Executive
Foretravel
Holiday Rambler
Landau
London-Aire
Rockwood
Sportscoach
Vogue

Truck Campers

(*Range:* $3,000 to $9,000)

Alaskan
Coachmen
Jayco
Shasta
Starcraft
Sunline
Viking

Comparison Shopping

Chapter 7

Whether you are shopping for a new or a used RV, a comparison spreadsheet can be a very useful tool. Believe us, before you are finished running around to all of those dealers out there and visiting the shows, you will be a very weary and confused shopper—a dangerous state in which to find yourself. You may reach a point at which any RV looks "good enough" and buy the next one you see just to get it over with. Then you are easy prey for a salesperson with any talent at all. With a spreadsheet, you can at least avoid the confusion and the feeling that you can't remember half of what you saw. Prepare the spreadsheet *before* you start your serious shopping.

Appendix B is a sample spreadsheet that you can use as a guide to make up your own. Use three sheets of letter-size paper and rule them as shown. Be certain that the lines coincide on the three sheets. Next, tape the sheets together side by side, using tape only on the back of the sheets. You should now have one wide spreadsheet that looks somewhat like the three sheets in appendix B.

On sheet A, record the basic information about the RVs

you want to compare, including the make, model, engine, and price. On sheet B, label each of the columns with the floor plan elements you have decided are most important to you. "Couch" is an example of these. Next, fill in the remainder of the columns on sheet B and as many as you need on sheet C with the options you would like to have. "Roof air" is an example.

What to do with this grid? Take it shopping with you, of course. The first RV you run across that you like, begin filling in the blanks. Be sure to check the boxes for floor plan elements and options. Make notes as to color, layout, or other features that will help you to recall the unit. It's a good idea to take down the dealer's phone number, too. After you have listed a few RVs, you will begin to appreciate the value of the spreadsheet in helping you to avoid making a major purchase error.

As for becoming weary of shopping, we can't do much about that other than to suggest dressing comfortably and wearing shoes with soles thick enough to insulate you from the gravel-covered sales lots you will encounter. If you get tired, stop for a breather or return to shop another day. Do not allow yourself to get to the exasperated stage. You will get to the point at which you continue shopping only to find the same things you looked at before. If you do, it is decision time. Review all of your records and make your choice before the fun goes out of it.

Other Tips

1. Look for the qualities you would seek in any salesperson—courtesy, honesty, and helpfulness. You are about to make a relatively large purchase, and you have the right to expect very good treatment. If you don't get it, find another dealer.

2. Before signing anything, read the document carefully.

3. Most of you will not have to be concerned with the following, but some of you please take note. When looking at RVs:

- If you turn something on, turn it back off.
- If you open something, close it.
- If you have children with you, control them.
- If you have a pet with you, clean up after it.
- And please, *do not* use the RV bathrooms.

Financing and Insurance

Chapter 8

Many banks will not finance a recreational vehicle. We believe that they do not understand how to deal with the product, because RV owners have demonstrated one of the best repayment records of all categories of loans.

Loan Provisions

The best policy is to do some telephone shopping first. Check rates at your own bank, and, if you belong to one, at your credit union. Check on rates and repayment terms at the dealer from whom you intend to buy. For example, can the loan be paid off early without penalty? When asking about rates, find out whether they are fixed or flexible. Flexible rates are usually tied to U.S. Treasury bill rates and can go up or down during the term of the loan. If you feel more comfortable with a fixed rate, then this is the way to go. Always have your rates quoted as the annual percentage rate, or APR. This is the *only* way to compare.

On a new unit, RV loans can run for longer terms than a car loan, in some cases for 8 or 10 years. But watch out—

interest on a long-term loan can add up to a formidable sum. Frequently, customers have asked us to sell their three-year-old motor home for them only to find that they owe more to the lending institution than the market value of the coach. As with the mortgage on a house, payments during those first few years of a long-term loan are mostly for interest, with very little going to principal payback. Meanwhile, the vehicle is depreciating faster than equity is building. So take the shortest loan you can that still allows you to make the payments without causing hardship. Many RV buyers choose to finance by means of a home equity loan. Again, check with your lending institution for rates and terms.

Life insurance and disability insurance for repayment of the loan are usually optional, but are recommended, especially in the higher loan amounts. Both types of insurance are available through the lending institution, and the premiums can be added to the payment.

Insurance

Insurance rates vary among insurance companies. Normally, you simply call the agent that insures your car and add the RV to the policy. Rates are usually lower, per dollar of insured value, than for a car. Because insurance companies know that RVs generally are not used as much as automobiles, the risk and rates are lower.

If you store an RV for long periods, such as over the winter, you might want to consider saving money by dropping all coverage except for fire, theft, and vandalism. Just be sure to reinstate full coverage before removing the vehicle from storage.

Ask your agent about additional coverage for contents. Remember that there is far more inside your RV than inside your car, so standard automobile content coverage will not be adequate.

Renting the RV to Others

If you rent your RV to others, be sure to notify your insurance agent. We promise that you will not like the rates, but this is only one of the many costs of renting out your RV. If you think you can recoup the cost of the vehicle, you had better give the issue further thought.

RV owners can rent out their units in two ways: privately (owner pays for upkeep, newspaper ads, etc.) or through placement in an RV dealer's rental fleet. The owner who handles the rentals will keep all of the profits. However, the unit is not as likely to be rented as often as it would be if it were placed in a dealer's rental fleet. Different dealers make rental arrangements with owners of the RVs in different ways. All take some percentage of the rental fee; some pay for maintenance out of their percentage, whereas others bill for this work. Usually, the owner pays the insurance cost, which can be a sizeable amount.

After each rental, the vehicle will need to be cleaned inside and out and any malfunctions checked and repaired. This must be done whether or not the unit is being rented through a dealer. If you live in a part of the country where winters are severe, the rental season is shorter, thus cutting income potential. Because rental customers shy away from high-mileage vehicles or older ones, the number of seasons the unit can be rented is limited.

The hidden or less obvious costs have already been stated. High mileage and shabby appearance will have a negative effect on the resale price of the unit—if you can sell it at all. Even though high mileage on a motor home may not be a big factor if the unit was well maintained, people are afraid of it. Note your own reactions if you shop for used units.

Before you decide to rent out your unit, do a lot of checking and ask questions. If you are working through a dealer, find out who pays for what and where the liabilities are. Talk to a tax expert about any claimed tax advantages. Then make your decision.

Taxes

Before you purchase an RV, seek the advice of your tax consultant. At the time of this writing, it is still possible to deduct all of the loan interest on an RV loan if it meets certain criteria. If the unit has a permanent cooking area, permanent sanitary facilities, and sleeping accommodations, it can be considered a second home. Other factors hinge on the extent that you use the vehicle. Some federal officials are advocating a change in this policy, so check with a tax specialist early in your shopping phase to see whether your plans for buying and using an RV will enable you to qualify for a tax deduction.

While you are thinking about taxes, it would be well to check into the taxes and/or license fees your state imposes on recreational vehicles. In some states, these fees can be substantial, especially as applied to motor homes.

What About Trade-Ins?

Chapter 9

Trading one vehicle for another can cost money—usually much more than you realize. A dealer is performing a service in taking your vehicle in trade, and services cost money. It *is* a service to you, because you will not need to shoulder the burden of trying to sell your own vehicle. Another thing to keep in mind is that dealers are in business to make a profit on each transaction. If they didn't, they would not be in business very long. The best way to understand how trades work is to put yourself in the dealer's shoes.

Imagine, for a moment, that you own an RV dealership. Being an astute businessperson, you have calculated the minimum profit that you are willing to accept for each unit on your lot. Knowing that most customers expect to bargain about price, you establish a "sticker price" that is higher than your cost plus minimum profit. This higher price also allows for negotiating flexibility when working with a trade.

Another quirk of human nature that you have learned is that the customer with a trade-in will usually expect a higher price for the unit being traded in than it is worth on the open market. So here come your first customers of the day. Being an experienced salesperson as well as a canny businessperson, one of the several opening questions you will ask is, "Do you have a trade-in?" The question might not be quite so direct, but you need to know this early on so that you'll know what you are up against when quoting prices. In this case, the answer is, "Yes, we have a car to trade." You are now aware of two factors that will affect the outcome of this transaction: One, the customers want to trade in an automobile on an RV, and two, you cannot afford to discount prices at this stage. As for trading the car, you also know that as an RV dealer, you would have little chance of reselling it from your lot because most people do not go to an RV dealer looking for cars. So your market is limited, and you do not want to have your money tied up in something that is not likely to sell in the near future. After all, you are buying that vehicle from the customers.

After some shopping, the customers show interest in a particular motor home. You casually ask, "Is there a figure that you have in mind that you expect to get for your car?" Sometimes you will get a direct answer to that question, but you have learned not to expect one. In this situation, the customers state a figure and ask for the cash difference between the value of the trade-in and the price of the motor home. It's now time to sharpen your pencil.

First, you call a car dealer friend to get an offer for taking this car off your hands. After being given a complete description of the car's condition, the dealer comes up with an offer. You also know that you have to make a profit from the motor home as well as a profit for the car when you *wholesale* it to your friend.

Get the picture? Your customers are expecting a figure

near *retail* for the car, but the most you can expect to sell it for is a *wholesale* figure. Where is the profit? In the sticker price, of course. On paper, you can show the customers a retail amount for the car by not discounting the motor home as much as you would have without a trade.

To clarify, here are some hypothetical numbers: The car is worth $6,000 retail and $5,000 wholesale (what your car dealer friend is willing to pay for it). You have decided that you need $500 profit from the sale of the trade-in; therefore, the most you can pay for the car is $4,500. Meanwhile, the motor home has a sticker price of $25,000—it cost you $21,000 and you want to make at least $1,000 profit on it. Add profit to cost, and the least you can accept for the motor home is $22,000 *without* a trade-in. The most you are willing to pay for the trade-in is $4,500. Subtracting $4,500 from $22,000 gives the amount ($17,500) that you need to get from your customers in addition to the car.

Now, working backward from $17,500, add $6,000, which is the amount the customers expect to receive for the car. The total is $23,500. You can still give your customers a $1,500 discount off of the sticker price. Here is how it will look to the customers:

Price of motor home	$25,000
Less discount	1,500
	$23,500
Less trade	6,000
Cash difference	$17,500

Here is how it will look to you, the dealer:

Received cash	$17,500
Value of car	4,500
	$22,000
Less cost of motor home	21,000
Profit	$ 1,000

However, the risk to you as a dealer rests with the sale of the car, because you still have to pay the manufacturer $21,000 for the motor home, and you received only $17,500 cash from the sale. Until the car is sold, not only will you not have made a profit, but you are out of pocket $3,500 (the difference between the $17,500 you received and the $21,000 that you owe). If you sell the car to your friend for $5,000, you will have made $1,000 on the motor home and $500 on the car. End result: The customers are happy, and you, the dealer, received the profit you needed to stay in business.

Now switch back to being yourself, the real customer. You are no longer a dealer.

Using the above example, what would have happened if you, the RV buyer, had price-shopped *without* a trade and got the discounted price of $22,000 for the motor home and sold the car privately for $6,000? The cash difference would have been $16,000, a savings of $1,500 to you. But there's a hitch. In the real world, you are not likely to give the private buyer of your car a warranty (unless you have a transferable warranty policy), and this is something that a car dealer usually will do. Also, the dealer will attract more people to the sales lot (volume of business) and is more likely to get the full retail price. However, even if you were willing to take $5,000 in a private sale transaction, you would still be ahead $500, less any advertising costs.

Why would anyone submit willingly to a trade-in? We are not psychologists, but we think we know the answer and have already stated it. Most people don't want to be inconvenienced by having to sell their own vehicle. Let's face it, it's easier to trade.

Dealers accept trade-ins to keep customers happy and to make an additional profit when reselling the traded vehicle. They are entitled to that profit because they are assuming the risks and investing their money in the trade-in.

So what does all of this mean to you? It means that you should always base your decisions on the cash difference, whether or not you decide to trade. You will have to assess your chances of selling your own vehicle and weigh the inconvenience of doing so against the convenience of trading. However, the cash difference can lean in your favor if you accept the risk of selling the vehicle yourself.

If you are still undecided about which way to go, ask dealers with whom you come in contact for the lowest price of the RV without a trade. Then check the wholesale "book" price of your potential trade vehicle. Do not expect a miraculous price for your trade at this point, because the dealer has already committed to a lower price on the RV. Instead, use a wholesale price for your trade. Then determine a fair retail price for your trade vehicle based on current "book" value. Do some figuring to determine the best approach to take. Keep in mind that you probably will not get full retail price when selling privately.

When comparing cash differences, make sure you compare apples with apples in terms of options and the quality of the product. Individual dealer policies governing trade-ins may vary from the example. Also, profit margin calculations in the example are hypothetical.

Things To Buy at the Camping Store

Chapter 10

Two self-stick bubble levels (see chapter 4).

Toilet chemicals. Follow the instructions on the package label and add the suggested amount to the sewer water holding tank. The chemicals are for breaking down accumulated solids.

Toilet paper. The special type made for camping is different from most supermarket varieties in that it is biodegradable to the maximum degree.

Sewer hose (10 or 20 feet), clamp and coupling. These items will be needed for dumping the holding tanks.

Water hose. Buy 50 feet of white plastic hose or the type that is self-contained on a reel. Because this hose will be used to fill your on-board fresh water tank and for connecting to a city water supply at the campsite, rubber hose is not suitable because it leaves a taste in the water. White hose looks more sanitary and is more visible, a safety concern.

Water pressure regulator. This device, which attaches between the water hose and the city water connection at the campsite, regulates the water pressure coming into your camper. Some campgrounds' water supply carries enough pressure to burst the plumbing in your vehicle. The regulator is cheap insurance. Use it *every time* you hook up.

"Y" connector for water hose. This is used if two campsites share a single outlet.

Hose nozzle. The pistol-grip type is preferable.

Shower curtain. You can forget this if one is supplied with your unit.

Stabilizing jacks. You will need these only if you have a trailer and if it is not equipped with them. They are used at the four corners of the frame to prevent the trailer from rocking on its springs as you move through it. They are adjustable and inexpensive. These are not leveling jacks.

Heavy-duty three-pronged extension cord. You will need 100 feet of it in a bright color.

Electric cord adapter. Adapts the large plug on the vehicle's power cord to a standard three-prong plug.

Fuses and bulbs. Be certain that the spare bulbs are the right type for both the interior and the exterior of your unit and that the fuses are the proper ones for the electrical panel.

A special refrigerator door lock. Although some refrigerators come equipped with a secure means of keeping the doors closed while traveling, some do not. Rather than take the chance of finding the contents of your refrigerator on the floor, we recommend this extra precaution.

Refrigerator shelf restrainers. A couple of types are available, and either of them will do the job. These devices prevent the contents of the refrigerator shelves from shifting or spilling while the RV is moving down the highway.

Refrigerator fan. These battery-operated fans are placed either inside the refrigerator to circulate the air for more even cooling or near the coils behind the refrigerator.

The latter type, which draws power from the RV's 12-volt battery rather than its own, needs to be wired in.

Plastic containers. To hold milk and other liquids in the refrigerator. Get the kind with screw tops.

Jack for changing tires. Most RVs are not equipped with jacks.

Wheel chocks. Used to prevent your vehicle from moving when parked.

Campground directory. Check the publication date and buy the most recent edition. Several different kinds are available from different publishers. Look through them to determine which one will best suit your needs. They usually contain everything you need to know about campgrounds throughout the country, including hookups available, whether pets are allowed, laundry facilities, and fees. A handy book to take along on your trip.

Igniter for stove. A safe device for lighting the stove and oven. Electronic or flint igniters are available. Either type will do the job.

Camp ax and saw. Handy for cutting and splitting small logs for that camp fire you don't want to miss.

Thin foam liner for cabinet shelves. Helps to keep things from sliding around when traveling. Also, it cuts down on rattles.

Sleeping bags. These help make the beds more comfortable, and they are much warmer than conventional blankets for those cool nights when you want to keep the furnace turned low.

Fire extinguisher. Although all new RVs are equipped with a fire extinguisher, make sure that you have one and that it works.

Read chapter 13 before you shop for camping supplies and equipment, as there are a couple of other items that you may want to add to this list.

Some Common-sense Safety Rules

Chapter 11

If you are an experienced camper, this chapter probably contains very little that you don't already know or that doesn't seem like common sense. However, a refresher won't hurt.

Before Starting Out

Before leaving on a trip, safety check the RV and tow vehicle or pickup truck if either is involved. Here is what to do.

1. Have all fluid levels checked on the chassis as well as the condition of all belts and hoses. Replace as necessary. Many campers take spares along as an added precaution.

2. Tune the engine if necessary. Have the compression checked, especially if you are headed into mountainous country. Good compression is important for engine braking on steep downgrades.

3. Have all suspension and steering components checked and repaired as needed.

4. Brakes should be checked for wear and operation.

This includes trailers! Don't skimp on this. If brakes are borderline, replace them. Remember that RVs are heavier than cars, and it takes more to stop them, especially in mountains. They must be in A-1 condition.

5. Repack the wheel bearings. Trailers should be attended to once a year—more frequently if you encounter very dusty road conditions.

6. If you own a trailer, have all hitch components checked for loose bolts or cracks.

7. Check tire condition and air pressure, including the spare. Replace any questionable tires—it is cheap life insurance. Trailer tires may be difficult to find while on the road. Most tire dealers do not stock them, so plan accordingly. Do not substitute smaller load capacities from the original equipment. Trailer tires are built differently from car or truck tires, so again, do not substitute.

8. Check the operation of all exterior lights and repair if needed.

9. Once the RV is loaded, take a trip to the closest truck weigh station and have the whole rig weighed to be sure you are not exceeding the gross vehicle weight limitations. If you are, trim down.

10. Have your RV dealer perform a propane check for leaks.

11. Before leaving on a major trip, take a camping weekend close to home for a shakedown of all systems to be certain everything is working properly.

12. Check the condition of fire extinguishers. If you have a trailer, keep one fire extinguisher in the tow vehicle and one in the trailer.

13. Check the contents of your first-aid kit. Replace items as needed.

14. A tip from Daniel McMahon, M.D., an emergency room physician friend: Take along a copy of the results of your latest physical exam. Include a copy of the EKG as

well as a list of any medications you are taking. Keep this information in a purse or glove compartment. Dr. McMahon says that having this information available can save time and lives in emergency situations.

On the Road

Once you hit the road, the following are some additional things to remember.

1. You are not driving a sports car. Recreational vehicles, as we have said earlier, carry a lot of weight and as a result build more momentum. They take longer to stop and cannot be maneuvered as well as an automobile in emergency situations. Leave plenty of space between your RV and the vehicle in front of you. Do not tailgate! Anticipate any turns or expressway ramps that you may want to take and slow down well in advance, using the appropriate turn indicators. Fast exits and RVs do not mix. Better to pass the turnoff than to wind up in a ditch or worse.

2. RVs are more prone to the negative effects of crosswinds or trucks passing than are cars because of the large surface area of the side of the unit. Be alert for passing trucks, and if strong, gusty wind conditions exist, slow down to a safer speed, or better yet, go find a place to park and wait it out.

3. If you are traveling through the mountains, remember the weight factor on steep descents. Slow down before any steep downgrade and shift to a low gear to let the engine provide some of the braking. Apply the brakes sparingly, but enough to keep the speed within the speed range of the gear in use. Do not pump the brakes. *Slower* is always better under these conditions. Because you have slowed down, you may discover a buildup of vehicles behind you. Don't get rattled. Take a scenic turnoff and let

them pass safely. Besides, this will give your brakes a chance to cool down. If you have been diligent about checking brake condition before the trip and use a low gear at a slow speed, you should have a safe journey.

4. If the unit should become disabled, place either road flares or reflective triangles at a good distance behind the RV. Turn on the emergency flashers. Move away from the vehicle and wait for help. If you have a citizens' band radio, use the emergency channel to summon help. Most police departments monitor this channel.

5. Before refueling the vehicle with gasoline or diesel fuel or having the propane tanks filled, check to see that all pilot lights in the RV have been extinguished.

6. Check tire pressure and brake, coolant, and oil levels frequently.

7. Because most states have a lower highway speed limit for RVs, be courteous and stay in the right-hand lane to allow automobiles safe passage.

8. Wear your seat belts.

9. When parked at a campsite, be sure to block the wheels fore and aft to prevent the RV from rolling. Do it even if the campsite appears level.

10. For the safety of your RV's water system while in winter storage, have your RV dealer winterize it for you or ask for instruction in the proper procedure for doing it yourself.

Some people are concerned about other kinds of safety when camping for the first time. They worry about everything from wild animals to wild people. As for animals, you have less to worry about in that department than you do from people. As for people, it has been our experience that you are safer camping than you are at home.

Whenever you camp in an unfamiliar part of the country, check with the ranger on duty or the campground owner about the local wildlife and get advice about precautions you should take. Remember that you are entering the animals' domain; respect them, and you will not have a problem. In most cases, the largest animal you can expect to encounter will be a squirrel or raccoon. But sometimes they are rabid, so stay away. Leave them alone and they will leave you alone. Most wildlife will avoid contact with people.

When in bear country, follow the rules and suggestions that the rangers will provide, and you will be perfectly safe. Don't pet or feed bears because they look friendly or cute and cuddly. This mistake could cost you your life. *Follow the rules.*

People are something else. Campers as a group are well known as the friendliest, most considerate people you will ever meet. They can be helpful almost to a fault, or they will leave you alone if you want to be left alone. But if you should need help, it is only a short walk or shout away. However, it takes all kinds to make a world. Don't leave attractions scattered about if you go on a hike. Put the grill and cooler away. Most campers will guard your property as if it were theirs, but there can always be exceptions. Use normal precautions, such as putting your valuables out of sight and locking windows and doors.

Private campground owners usually live on the property, and park service personnel patrol public campgrounds. As we said, you and your property are probably safer on a campground than at home. However, people *outside* the campground can be a problem. Some would like to make off with your vehicle or rummage through it while you are sightseeing. A good alarm system will prevent all but the most professional attempts. You have the same problem if you leave a desirable car unattended.

For your sake and that of others, make certain that your

camp fire or charcoal grill is completely extinguished before you leave the premises. Be aware of local fire conditions and obey the rules.

Keep track of your kids. Educate them to respond to a whistle or buy a beeper system or walkie-talkie to keep in touch with them.

Checklist of Things To Take on Your Trip

Chapter 12

You will need to plan your first camping trip carefully, paying special attention to your personal needs, cooking needs, and vehicle requirements. We have made this checklist for you to consider as you make your own list. Plan to take only the supplies you know that you will need. Most of us tend to take more than we will use when we start out on a camping trip.

Vehicle Supplies

These items will be useful both for safety and convenience:

- ☐ Toilet chemicals
- ☐ Toilet paper
- ☐ Shelf retainers for refrigerators
- ☐ Baking soda for refrigerator
- ☐ Igniter for stove
- ☐ Television set

- ☐ Audio tapes for tape player
- ☐ Flashlight and extra batteries
- ☐ Extra motor oil, fan belt, and radiator hoses
- ☐ Tools (make a list)
- ☐ Spare fuses and bulbs
- ☐ Fire extinguisher
- ☐ Ramps for leveling
- ☐ Wheel chocks
- ☐ Road flares
- ☐ Water hose and hose nozzle. (drinking and garden, 50 feet each)
- ☐ Sewer hose
- ☐ Water pressure regulator and "Y" connector
- ☐ Extension cord and adaptor
- ☐ Tire gauge
- ☐ Stabilizing jacks (if you will be pulling a trailer)

Personal Care and Clothing Supplies

Plan your wardrobe carefully to include items for anticipated temperature extremes. In some parts of the country, even summer nights can be chilly and damp. You won't want to leave your camp fire to sit inside because of the lack of warm clothing.

- ☐ Bathroom supplies (razor, toothbrushes, etc.)
- ☐ Towels and washcloths
- ☐ Footwear for public showers
- ☐ Beach towels and swimwear

- ☐ Clothes (make a list)
- ☐ Hiking boots
- ☐ Rain gear
- ☐ Sewing kit
- ☐ Iron and portable ironing board
- ☐ Clothesline and pins
- ☐ Clothes-washing detergent, fabric softener, and liquid bleach
- ☐ Insect repellent and snakebite kits
- ☐ First-aid kit
- ☐ Sunglasses
- ☐ Sun protection head covering

Cooking Supplies

Plan your menus in detail and take only the quantity of food you will need. Avoid glass containers; transfer what you need to an unbreakable container such as plastic. Line cooking pots with foil and consider using disposable plates and drinking cups. Simple and easy cleanup leaves more time for fun.

- ☐ Collapsible 5-gallon jug for emergency water needs
- ☐ Plates, cups, and bowls—plastic or paper
- ☐ Eating utensils
- ☐ Cooking utensils
- ☐ Plastic containers and lids for food
- ☐ Coffeepot
- ☐ Pots and pans
- ☐ Toaster and other favorite appliances

- ☐ Can opener
- ☐ Charcoal grill and utensils
- ☐ Charcoal and lighter fluid
- ☐ Dish towels and pot holders
- ☐ Large plastic garbage bags
- ☐ Paper towels
- ☐ Heavy-duty aluminum foil
- ☐ Matches
- ☐ Broom
- ☐ Food for guests and pets (include staples, and make a list)
- ☐ Dish detergent and scouring pads

Sleeping Gear

If your vehicle does not have a furnace, be sure to take warm bedding. A sleeping bag can make a big difference in your comfort at any season. Practical nightwear and a robe can be handy if you have to leave your camper at night for any reason. With warm sleepwear and bedding, you'll sleep better than ever before—all that fresh air.

- ☐ Sleeping bags with liners or sheet protectors or . . .
- ☐ Sheets and warm blankets
- ☐ Pillows and cases
- ☐ Sleepwear and robe

Other Useful Items

Depending on the size, ages, and interests of your camping partners, you will want to consider these items for recreation and for the type of campground destination:

- ☐ Portable radio
- ☐ Games for rainy days
- ☐ Reading material
- ☐ Paper and pencils
- ☐ Wasp and hornet spray
- ☐ Hunting knife, camp ax, and saw
- ☐ Wood and kindling for a fire
- ☐ Lawn chairs
- ☐ Fishing and hunting gear
- ☐ Maps
- ☐ Campground directory
- ☐ Camera and film
- ☐ Pet leashes and/or chains (required by most campgrounds)

Once you have made several camping trips of various duration, you will develop your own checklists. Each time, the preparation gets easier and faster, and you'll discover that "less is better."

Some Pointers About Camping in an RV

Chapter 13

The purpose of this chapter is to provide some tips to make life a little easier at the campsite—when you arrive, while you are there, and as you prepare to leave.

Arrival

Parking. After you have your preferred or assigned camping spot, the first task is going to be parking the unit. If others are with you, have them inspect the area in which you will be parking. Watch for any ground obstructions or potential tree limb problems from above. Large dead branches have been known to come down in a storm and put nasty holes in the roofs of RVs. With a companion positioned at the rear of the unit and in full view of your mirror, back in slowly, relying on the helper for directions. The helper needs to watch not only where you are going but also for low-hanging branches. Don't assume that your unit is tough enough to overcome the strength of a tree branch. If you do, a dent at the roofline or a crack in the fiberglass can almost be guaranteed.

If you are parking a trailer, place your hand at the *bottom* of the steering wheel when backing up. This way, the trailer will go in the direction your hand moves. This method is easier for most people and less confusing than having to remember to turn the wheel in the opposite direction of the way you want the trailer to go, as is the case when placing your hand on top of the wheel.

Your first couple of times out, get to the campground while plenty of daylight is left so that you can be settled in before dark. It's much easier on tempers to accomplish the parking and setup without having to resort to flashlights. You will be safer from tree limbs too, because they're harder to spot at night.

Check power source. Make certain that the 110-volt outlet at the campsite is operational before doing any additional setup. Plug-in testers are available at camping or hardware stores, but you can simply use a night-light to see if you have power. Most campsite electrical boxes are equipped with circuit breakers, so be sure it is switched to the "on" position for your test.

Level your camper. Refer to chapter 4 if you need a refresher. Set stabilizing jacks at all four corners of the frame if you have a trailer.

Electrical hookup. Check to see that the circuit breaker in the electrical box at the campsite is in the "off" position. Plug into the 110-volt outlet, *always* using the bright-colored extension cord you bought at the camping store. The 25 feet of cord usually supplied with most vehicles is black and harder to see, thus increasing the chances of tripping over it. Any excess cord should be neatly coiled and placed close to the vehicle. Again, this is to prevent accidents. When connecting the cord to the outlet, keep your fingers out of the outlets and away from the prongs. Use caution, especially if it is raining. Make sure that the circuit breaker in the electrical box is in the "on" position after connecting.

Ask a companion to plug a 110-volt appliance into a 110-volt outlet in the vehicle to see if it operates. If it does not, check the breakers and fuses in the vehicle.

Water hookup. If you have a water hookup, get out the white hose. Connect one end to the outside fixture on your RV labeled "city water." Attach the pressure regulator to the other end of the hose. Before attaching the hose to the campground water faucet, turn it on and let it run for a minute to clear the pipes of any accumulated rust or stagnant water. Turn it off and attach the hose. Make sure that all the faucets in your vehicle are turned off before turning on the water supply. After the water is on, turn on the faucets inside the vehicle very slowly so that the air can escape from the lines.

Sewer hookup. This section assumes that your RV is equipped with holding tanks. There are usually two—one a gray water tank and the other a sewer tank. Gray water is the wastewater collected from the kitchen sink and usually the shower. Sewer waste comes from the toilet and usually the bathroom sink. Both tanks are normally piped to a common drain. Some units may have a separate outlet for each tank, but this is *not* a desirable feature. Each tank is equipped with a "slide valve" that is operated from the outside of the RV (see figure 13-1). When open, the valve allows the contents of the holding tank to flow through the outlet and attached sewer hose and into the sewer at the campsite or dump station. Some older units have only one holding tank that must handle all of the wastewater. One tank usually means smaller capacity compared with a two-tank system, therefore demanding more frequent emptying (dumping).

Keep in mind that most private campgrounds and many

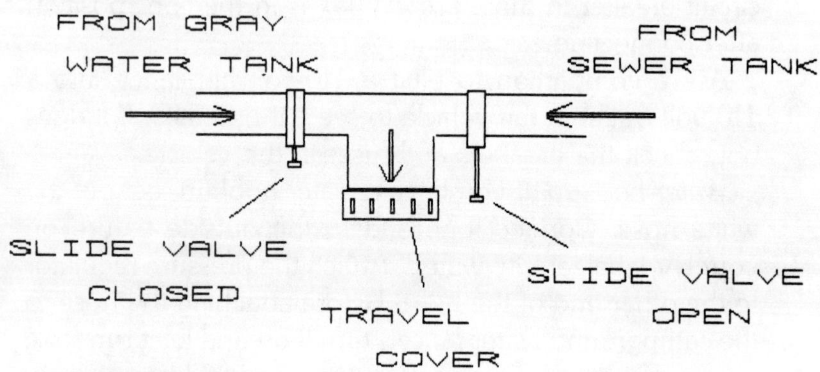

Top view of common drain connected to two holding tanks, with slide valve shown in open and closed positions. (figure 13-1)

public ones have good, clean restrooms. You may choose not to use the toilet in your RV and use the campground facilities instead. We have had many used units on our lot in which neither the toilet nor shower has ever been used. In any event, it will still be necessary either to hook up to a sewer connection at the campsite or to dump the holding tanks periodically in order to rid the tank of accumulated soapy dishwater, bathroom sink water, or shower water.

If your RV has holding tanks and the site has a sewer hookup, attend to this matter next. Making sure that the holding tank valves are closed, remove the travel cover from the tank outlet. Remove the sewer hose from storage and attach the end with the coupler to your unit's holding tank outlet. Take the other end of the hose and insert it into the sewer pipe at the site. *Do not* open the RV holding tank valves yet. The sewer connection is used only after waste has collected in the holding tanks. Collected waste in the tanks creates a positive flow from them when it is time to "dump."

When to dump depends upon how much waste you

create and how fast the tanks fill. When flushing the toilet, do not skimp on the amount of water you use and make sure that there are chemicals in the holding tank. Most campers will dump the tanks before leaving the campsite after a weekend or after three or four days of traveling.

A Word About Pets

Most animals love to go camping. If you decide to take them along, keep them on a leash or chain at all times. If they are dogs that bark excessively, leave them at home in the care of a friend, or board them when you travel. When leaving pets alone in the RV, make sure they have plenty of ventilation and water.

Flea and tick collars are recommended, along with frequent checks for these pests. If dog-walking areas are provided, use them. If the dog creates a mess at your campsite, clean it up.

Before Leaving the Campsite

After camping, dump the contents of the holding tanks. This task, like leveling, takes longer to read about than it does to perform.

Attach the sewer hose as described on the previous page. Dump the sewer tank first. Open the valve and let the tank drain. Then, from inside the coach, rinse the tank using a high pressure wand and hose attached to your bathroom sink. You lower the wand into the tank through the valve in the toilet. (The hose and wand are available at most camping stores — add them to your shopping list.) After rinsing, close the outside sewer valve. Now open the gray water tank valve and allow that tank to empty. This will help to flush any remaining sewage from the hose. Detach the hose from the RV end first.

Using a separate garden hose (not your white one), connect it to the faucet at the site and rinse the inside of the hose, letting the water run into the sewer. Remove the sewer hose from the sewer and store it. Replace the travel cover on the tank outlet. Inside again, add chemicals to the sewer holding tank, along with enough water to cover the bottom of the tank. That's it!

You will find that most campgrounds, as well as some highway rest areas, have dump stations, so if you did not have a sewer hookup at your campsite, you can use these. In addition to the sewer connection, these facilities usually are equipped with a water hose for cleanup. Do not dump the holding tanks into a stream, river, or lake, or anywhere along the highway. It is illegal.

Drain the freshwater tank. If you won't be using any water for cooking on the way home, now is a good time to drain the freshwater tank. Make certain that the water pump is turned off. Open the drain for the tank and close it when empty to prevent road dirt from entering the tank.

Clean up the campsite. Appendix C is a reminder checklist for departure from the campground. We hope you have a great time on your camping trip.

The Last Word

Chapter 14

Well, here we are at the final chapter. What we have written so far may seem overwhelming if you have never owned an RV, but we hope that it has not discouraged you from moving ahead. A recreational vehicle offers a way of life that, once in your blood, is not easily removed. As we said earlier, many tasks for which we have given instructions really do take longer to read about than to perform. After you have done them a couple of times, they will become routine. We also said that you will find more help from your fellow campers than you can ever use. The first time you have to set up or level your RV or the first time you have to dump the holding tanks, ask a neighbor for advice. You'll get it, along with shortcuts that will make the next time easier. We still learn things from other campers even though we've been camping for years. That's part of the pleasure of the RV life.

Chances are good that you will share your campground with quite a number of people from other countries. It is a popular way for them to tour the United States and makes for some interesting evenings for you around the camp fire.

There are many camping clubs, some private and some sponsored by the manufacturer of your RV. Most offer group tours and usually a yearly national get-together. If you like group activities, the opportunities are there.

Many state and national campgrounds offer nature films, programs, and discussions in the evening hours, weather permitting. Many private campgrounds have swimming pools, game rooms, play areas for the kids, and camping supply stores. Activities can range from horseshoes to ballroom or square dancing. But if you just want peace and quiet, it's there, too.

If you are an experienced camper poring over these pages, we hope that you have gained some insight into the workings of the RV business and that, as a result, you will be able to do a better job of shopping when you find yourself ready for your next RV. We know that there will be a next one—there almost always is.

If first-timers find themselves getting the shopping blues, then reread the first chapter, do some dreaming, and put yourself squarely in the picture. Camping is fun. It's the ultimate getaway. Don't miss out!

Appendix A

Checklist for Evaluating a Used RV

YEAR, MAKE, MODEL ______________________________
SELLER _______________________________________
COMPREHENSIVE INSPECTION

APPEARANCE AND EQUIPMENT EST. REPAIR COST

- ☐ Overall condition
- ☐ Exterior roof—tar? sags? holes? tears?
- ☐ Body panels—damage? repairs? pitting?
- ☐ Exterior storage compartments—water damage?
- ☐ Tire condition—sidewalls? tread wear?
- ☐ Spare tire condition
- ☐ Bumper condition
- ☐ Exterior lights—head? brake? turn? markers?
- ☐ Holding tanks and pipes—leaks? cracks?

- ☐ Hot water heater—pilot? main burner?
- ☐ Generator—starting? smooth operation? oil?
- ☐ Awning—tears? bent arms? spring loading?
- ☐ Interior ceiling—sags? soft spots? stains?
- ☐ Interior floor—soft? stains?
- ☐ Interior walls—soft under windows? at ceiling?
- ☐ Upholstery condition
- ☐ Carpet condition
- ☐ Cabinet & hardware condition
- ☐ Bathroom condition—sink/tub cracks? toilet cracks?
- ☐ Mattress condition
- ☐ Propane leaks (sniff the air with propane on)
- ☐ Furnace—ignition? blower?
- ☐ Stove/oven—ignition? oven pilot and burner?
- ☐ Converter (Do lights brighten when plugged in?)
- ☐ Refrigerator operation
- ☐ Roof air-conditioning
- ☐ Water pump—leaks? continuous running?
- ☐ Toilet flushing
- ☐ Hot water heater tank—leaks?
- ☐ Fresh water tank—leaks?
- ☐ Visible plumbing—leaks?
- ☐ City water hookup—leaks?
- ☐ Monitor panel—lights?
- ☐ TV antenna—crank operation? water leaks?
- ☐ TV operation
- ☐ Microwave oven—working?

- ☐ Stereo—all functions?
- ☐ Power entry step (Does it fully extend and retract?)
- ☐ Power leveling jacks (Do they fully extend and retract?)
- ☐ Ice maker—working?
- ☐ Washer and dryer—working?

DASHBOARD COMPONENTS

- ☐ Stereo—all functions?
- ☐ Air-conditioning—working?
- ☐ Heater—all functions?
- ☐ Instruments—gas? charge? oil?
- ☐ Windshield wipers and washers—working?
- ☐ Dash lights—working?

MECHANIC CHECKS/TEST DRIVE

- ☐ Dip sticks—engine oil? transmission oil?
- ☐ Engine—misfiring? pinging? valve noise? etc.
- ☐ Transmission—slipping? noise? shifting?
- ☐ Rear end—leaks? noise?
- ☐ Steering
- ☐ Exhaust
- ☐ Suspension
- ☐ Tire wear
- ☐ Batteries
- ☐ Speedometer
- ☐ Cruise control
- ☐ Frame—visible damage?

■ *Recreational Vehicles*

- ☐ Exterior body
- ☐ Additional propane leak test by RV dealer

OTHER ITEMS TO CHECK

- ☐ Warranty?
- ☐ Read purchase contract before signing

Appendix B

Spreadsheet for Comparison Shopping

Sheet A

MAKE	YEAR	LGTH	ENGINE	PRICE	DEALER	SALESPERSON

Sheet B

COUCH										ROOF AIR							

Sheet C

										NOTES

Appendix C

Campground Departure Checklist

- ☐ Lower TV antenna.
- ☐ Stow all loose gear and appliances.
- ☐ Check all cabinet doors for tight closure.
- ☐ Switch refrigerator to gas or battery for travel.
- ☐ Secure contents of refrigerator and lock it.
- ☐ Raise and secure awning.
- ☐ Turn off all interior lights.
- ☐ Turn off stove and furnace.
- ☐ Unplug and stow electric cords.
- ☐ Disconnect water hose and stow.
- ☐ Dump holding tanks if used for several days.
- ☐ Clean and stow sewer hose.
- ☐ Remove and stow stabilizing jacks.
- ☐ Remove and stow wheel chocks.

- ☐ Move vehicle from leveling boards and stow boards.
- ☐ Secure all outside storage doors.
- ☐ Close all windows and roof vents.
- ☐ Place entry step in travel position.
- ☐ Lock entry doors.
- ☐ If RV is a trailer, double check hitch.
- ☐ Check operation of all outside lights.
- ☐ Clean up campsite.

When you get home, clean RV inside and out. Prop open refrigerator or icebox door to prevent mildew. If you own a fold-down or pop-up roof, open it to allow any moisture to dry out.

Index

A

B

C

D

E

F

S

T

U

V

W